I0431212

THE RIDDLE OF THE UNIVERSE IS SOLVED!
THE THEORY OF EVERYTHING DISCOVERED!

NOBEL PRIZE CONSIDERATIONS:

1. THE UNIVERSE CONSISTS OF MANY "SPECIAL NUCLEAR REACTIONS," AND IS IN THE STATE OF "PERPETUAL MOTION." "THE UNIVERSAL SYSTEM" IS IN EQUILIBRIUM – IT DOES NOT LOSE OR NEED MORE ENERGY OR MATTER TO FUNCTION FOREVER.
2. THERE IS A WORKING BLACK HOLE INSIDE THE EARTH, AND IS IN TROUBLE RIGHT NOW WHEN OUR LIFES DEPEND ON IT 100%.
3. THE HIGHEST TEMPERATURE IN THE EARTH IS ABOUT 3500 KM UNDER THE SURFACE AND NOT IN THE CENTER OF THE PLANET, AND IS CAUSED BY THE "SPECIAL NUCLEAR REACTION."
4. IN THIS BOOK THE REAL "GRAVITY" THAT "COMMANDS" THE UNUVERSE HAS BEEN DISCOVERED – "UNIVERSAL COMMANDING INFLUENCE." NEWTON'S GRAVITY APLLIES ONLY FOR FALLING APPLES, AND NOT EVEN FOR FALLING IRON. IT HAS NOTHING TO DO WITH PLANETS ORBITING THE SUN, OR MOONS ORBITING PLANETS, AND STARS STAYING IN GALAXIES.
5. ONE OF THE MANY "BIG UNIVERSAL AND LOCAL EXPLOSIONS" THAT CREATED THE LOCAL GROUP OF GALAXIES WAS MISTAKEN FOR SOME SORT OF "BIG BANG." NO "BANG" AS BIG AS IT MIGHT BE COULD CREATE THE ENTIRE UNIVERSE WHICH SIZE IS UNKNOWN.
6. THE ENERGY IN THE UNIVERSE IS "FREE;" IT IS NEITHER CREATED NOR LOST. THERE IS NO SUCH CONVERSION OF MATTER TO ENERGY OR VICE VERSA. "E = MC2" IS COMPLETE NONSENSE AND SHOULD BE TRASHED RIGHT NOW.
7. THE UNIVERSE CONSISTS OF THREE MATTERS: "MATTER # 1" BASIC BUILDING PARTICLES OF THE ATOM. TOGETHER WITH "MATTER # 3" ~~WHICH~~ IS 97% OF THE TOTAL, AND IS FOUND ONLY INSIDE ALL BLACK HOLES; "MATTER # 2" ABOUT 3% OF ATOMS AND MOLECULES; ~~AND~~ "COMPACTED" "MATTER # 1" IS CALLED "MATTER # 3" AND IS FOUND ~~LESS THAN 1% FOUND~~ IN NEUTRON STARS AND WHITE DWARFS ~~CALLED "MATTER # 3"~~. ~~THE PERCENTIGES VEARY.~~

8. "BLACK HOLES" SHOULD BE CALLED "BLACK SPHERES." SCIENTISTS TODAY KNOW ONLY "MATTER # 2," BECAUSE IT IS ON THE SURFACE OF THE EARTH, BUT "MATTER # 1" IS INSIDE OF EACH STAR AND IN THE CENTER OF <u>ANY "LIVE"</u> ~~OUR~~ PLANET OR MOON<u>'S</u> ~~"SOLID CORE."~~
9. MATTER # 2 IS CREATED ONLY TEMPORARILY <u>(5 BILLION YEARS?)</u>, AND THAT IS ALL PLANETS, MOONS, AND HUMANS, BUT MATTER # 1 <u>TOGETHER WITH</u>~~AND~~ ITS ENERGY "LIVE<u>S</u>" FOREVER.
~~IN THE UNIVERSE MATTER AND ENERGY ARE TWO SEPRATE ENTITIES, BUT ARE UNSEPERABLE AND LIVE FOREVER.~~
10. THERE IS NOT A SINGLE ATOM OF HYDROGEN GAS IN ANY OF THE STARS!
11. MATTER NUMBER ONE GOES THROUGH ROUGHLY "20 BILLION-YEAR LONG "~~LOCAL~~ UNIVERSAL CYCLES" WHICH IS SEPERATE FOR EACH CLUSTER OF GALAXIES AND IS COORDINATED BY "THE GRAND UNIVERSAL DESIGN."
12. THE CONSTANT REPETITION OF THESE "~~LOCAL~~ UNIVERSAL CYCLES" DEFINES THE "EVERLASTING" EXISTANCE OF THE ENTIRE UNIVERSE.
13. OUR MOON HAS NO RIGHT TO GO 3.3 CENTIMETERS AWAY FROM THE EARTH EACH YEAR. LET US FIND OUT HOW THIS "SCIENTIFIC" BALONEY WAS "DISCOVERED?"
14. THE LOCAL "CRUNCH" DICTATED BY THE 20 BILLION YEARS "CYCLE" FOR THE LOCAL GROUP TO WHICH OUR GALAXY BELONGS IS AT HAND RIGHT NOW.
15. THE DOPPLER SHIFT AND HUBBLE LAW OF EXPANDING THE UNIVERSE ARE ABSOLUTE NONSENSE, AND SHOUD BE SUBSTITUTED WITH THE REAL "RED-SHIFT CONSTANT OF DISTANCE" INVENTED BY THE AUTHER: "THE FARTHER THEY ARE, THE MORE RED SHIFT THEY ACQUIRE, BECAUSE THROUGH MORE ~~PARTICLES~~<u>DUST AND PARTICLES</u> THE LIGHT HAS TO GO <u>THROUGH</u> ~~THROUGH~~."
16. EINSTEIN'S THEORIES OF RELATIVITY ARE ABSOLUTE AND COMPLETE NONSENSE AND SHOULD BE TRASHED 100% RIGHT NOW!
17. THE "GREAT MATHEMATICIAN" ISAAC NEWTON FAILED TO UNDERSTAND KEPLER'S THIRD LAW, AND MADE AN ENORMOUS ERROR RESULTING IN THE MISCALCULATION THE MASS OF OUR SUN!!!
18. ALL STARS ARE BLACK SPHERS ~~AND~~ RELEASING BASIC BUILDING PARTICLES OF THE ATOM (BBPA), SO DOES OUR SUN,

AND THAT WHAT THE SOLAR WIND IS. THE HEAT FROM THE SUN DOES NOT COME FROM INSIDE, BUT IS CREATED OUTSIDE THROUGH THE "SPECIAL NUCLEAR" REACTION WHERE SELF-CREATION OF PLASMA, ATOMS, AND MOLECULES IS HAPPENING. THIS IS A NUCLEAR REACTION WHICH SCIENTISTS DO NOT KNOW AND UNDERSTAND YET.

19. ALL PLANETS, MOONS, AND SPHERICAL ASTEROIDS ARE CREATED INITIALLY BY BLACK SPHERES INSIDE THEM, AND SOME OF THEM FUNCTION RIGHT NOW, BUT SOME ARE "DEAD." EARTH'S BLACK SPHERE IS ALIVE, BUT THE MOON'S IS "DEAD."

20. FROM THE FOUR GAS GIANTS WHICH ARE BROWN DWARFS JUPITER IS THE WEAKEST, AND NEPTUNE IS THE SRONGEST.

21. THERE IS NO ICE INSIDE ANY OF THE GAS GIANTS - ALL OF THEM HAVE VERY POWERFUL LIVE BLACK SPHERES WORKING RIGHT NOW.

22. THE HEAT INSIDE OF THE "LIVE" PLANETS AND MOONS COMES FROM THE SELF-CREATION OF BIGGER PARTICLES (PROTONS), ATOMS, AND MOLECULES AS THE BASIC BUILDING PARTICLES WHICH ARE RELEASED FROM THEIR BLACK SPHERES LOCATED IN THE CENTER OF THESE CELESTIAL BODIES.

23. LIVE VOLCANOES ON ANY CELESTIAL BODY ARE INDICATION FOR WORKING BLACK SPHERES INSIDE THEM. IF EARTH'S VOLCANOES STOP WORKING, WE WOULD DIE WITH THEM OR BECOME SPACE ALIENS IF PREPARED.

24. VENUS HAS A BIGGER BLACK SPHERE THAN THE EARTH, AND COULD BE QUALIFIED AS A BROWN DWARF (MY CLASSIFICATION).

24. ~~THE EARTH HAS THE BIGGEST POSSIBLE BLACK SPHERE FOR A CREATES EXESSIVE HEAT AND RADIATION, AND SURFACE WATER AND LIVING ATMOSPHERE ARE ABSOLUTELY IMPOSSIBILE TO DJUPITER'S MOON IO IS A BROWN DWARF, AND IS MORE IN OUR GALAXY EXEPT OTHER "SOLAR SYSTEMS," THERE~~

25. SATURN IS TOO POWERFUL BROWN DWARF TO HAVE A MAGNETIC FIELD, THE WEAKER ONE - JUPITER HAS THE STRONGEST POSSIBLE ONE FOR A BROWN DWARF.

26. THE SUN MAKES THE PLANETS ROTATE AROUND THEIR AXES, AND THIS IS PART OF ITS "COMMANDING INFLUENCE" IN THE SOLAR SYSTEM.

27. MATTER AND ENERGY ARE THE TWO SIDES OF THE SAME COIN AND BOTH ARE INDESTRUCTABLE AND INSEPARABLE.

28. THE 4 PLANETS AND THE 4 BROWN DWARFS TOGETHER WITH THE SUN IN OUR SOLAR SYSTEM DO NOT ALLOW MOONS TO ROTATE AROUND THEIR AXES, AND THAT IS WHY THE MOONS DO NOT HAVE "MOONLETS."
29. CURRENT FOREST FIRES, GLOBAL HAVOC, AND WARMING ARE CAUSED BY A POWERFUL "INFLUENCE" OF ANOTHER CELESTIAL BODY, WHICH COULD BE AN~~THE~~ "INTRUDER" ~~WHICH MIGHT~~THAT COULD BE A BROWN OR A WHITE DWARF THAT CAME INSIDE THE SOLAR SYSTEM AND DAMAGE OUR MAGNETIC FIELD THROUGH DISPLACING OUR BLACK SPHERE, AND IT IS NOT OVER YET.
30. BLACK SPHERES CHANGE THEIR DENSITIES, THEREFORE THEIR MASS. THEIR SIZES REMAIN THE SAME REGARDLESS OF THEIR CONSTANTLY DIMINISHING CONTENT OF "BBPA."
31. THE SIZE OF THE SUN MIGHT VEARY A BIT, BUT IN A LONG RUN REMAINS THE SAME.
32. THERE IS NO DIFFERENCE IN CONTENT AND STRUCTURE BETWEEN PLANETS AND MOONS – THEY ARE ONE AND THE SAME. IO AND TITAN ARE EXEMPTIONS – THEY ARE BROWN DWARFS.
33. THE UNIVERSE IS ONE GIANT SYSTEM CONSISTING OF SMALLER AND SMALLER SYSTEMS IN IT. FIRST SYSTEM IS THE ENTIRE UNIVERSE; SECOND ARE ALL CLUSTER OF GALAXIES; THIRD IS A "GROUP OF GALAXIES;" FOURTH IS A SINGLE GALAXY; FIFTH ARE THE "SOLAR SYSTEMS;" AND FINAL SIXTH AND THE SMALLEST ONE IS A PLANET (BROWN DWARF) AND MOONS SYSTEMS. EVERYTHING IN THESE SYSTEMS IS COORDINATED, AND NOTHING MAJOR IS ALLOWED TO HAPPEN BY CHANCE.
34. CURRENTLY THE SCIENTISTS DO NOT KNOW HOW LONG THE EARTH, THE SUN, AND THE GALAXY WOULD LAST, DUE TO LACK OF KNOWLEDGE.
35. ALL STARS IN OUR GALAXY AND THE SUN ARE APPROXIMATELY 13.8 YEAR OLD.
36. 13.8 BILLION YEARS AGO THE MILKY WAY (SPIRAL GALAXY), MIGHT HAVE BEGAN ITS EXISTANCE AS DIFFERERNT GALAXY, AND HAS BECOME SPIRAL IN THE LAST 5 BILLION YEARS, AND IN THE FUTURE WOULD BE PART OF AN ENORMOUS ELLIPTICAL ONE.
37. EVERYTHING IN THE UNIVERSE IS COMMANDED BY THE BASIC BUILDING PARTICLE OF THE ATOM REGARDLESS WHERE THEY ARE – INSIDE A BLACK SPHERE OR IN THE REGULAR ATOMS.

THEY ARE INDESTRUCTABLE AND "EXISTS" FOREVER LEADING A "DYNAMIC EXISTENCE." THEY CARRY ENERGY WICH IS NEVER LOST OR SEPERATED FROM THEM.
38. ALL CELESTIAL BODIES POSESS "INFLUENCE." THE "GRAVITY" IS PART OF IT, BUT DOES NOT EXTEND ~~FARTHER~~ABOVE TH~~AN~~E ROCHE'S LIMIT. THE "INFLUENCE" THAT EXTENDS ABOVE THE ROACH'S LIMIT~~FATHER~~ IS DIFFERENT!
39. DOES THE EARTH INCREASES IN SIZE RIGHT NOW? YES, THE NEWLY CREATED MATTER PUSHES LAVA THROUGH VOLCANOES AND OTHER CRACKS IN THE CRUST.
40. IF MARS HAD AN ATMOSPHERE BEFORE, THEN IT MIGHT HAVE DEPOSITS OF GAS, OIL, OR COAL. ALTHOUGH THEY COULD NOT BE IN THE "THARSIS VALLEY," SO THEY SHOULD BE FAR FROM IT OR ON THE OTHER SIDE OF THE PLANET.
41. JUPITER MOON IO IS MORE POWERFUL THAN THE EARTH; ~~THEREFORE~~THEREFORE, THE EARTH COULD HAVE BEEN MOON TO ANY OF THE 4 GAS GIANTS.
42. IN OUR GALAXY EXCEPT OTHER "SOLAR SYSTEMS," THERE SHOULD BE MANY "BROWN DWARF" SYSTEMS, WHERE WE DO NOT SEE NEITHER THE BROWN DWARFS AT THE CENTER OR THE PLANETS GOING AROUND THEM.
43. THE EARTH HAS THE BIGGEST POSSIBLE BLACK SPHERE FOR A PLANET IN THE ENTIRE UNIVERSE, AND HAS THE BIGGEST POSSIBILE SIZE FOR A ROCKY PLANET, IF A "PLANET" HAS A BIGGER OUTER SIZE THAN THE EARTH, IT IS A "BROWN DWARF" (MY CLASSIFICATION).
44. WHAT WE REFFER AS ENERGY IN THE UNIVERSE IS PARTICLES AND ENERGY – THEY ARE INSEPERABLE!
45. THE "ORDER" IN THE UNIVERSE IS "ESTABLISHED" BY THE "INFLUENCE" OF THE LARGER BLACK SPHERES UPON THE SMALLER.
46. IT HAVE BEN STATED THAT JUPITER OR NEPTUNE COULD "THROW" "MOONS" OUT OF THE SOLAR SYSTEM. THIS IS IMPOSSIBLE! THERE IS ONLY ONE "BOSS" – THE SUN.

QUOTATIONS FROM "UNIVERSE" SMITHSONIAN 2020

All textbooks reflect the current knowledge in the science of Astronomy all over the world, and the current deceptions as well. Basically I would have picked up any text book, but just happens that this was the book available at that time in a bookstore. I apologize to the writers of this book. I do not mean to offend you.

SCIENCE AND SCIENTISTS

In this book I am using quite derogatory language about the scientists in general and some particular historical individuals. I have great respect for all scientists in general, but the "job" of scientific inspector requires finding the "scientific truth."

Recently, already as a senior citizen I become interested in the science of Astronomy. In some other sciences the scientific results are more real, but in the science of Astronomy wild speculations are something common and widely accepted. I am trying to excuse myself about my derogatory remarks used here. I would characterize my remarks as border-line in decency, but keep in mind that my task is Herculean, the entire human race has been grossly deceived by some "scientists," and I am chosen to save the world and the human race from these enormous deceptions. My measure of things is the "scientific truth," and my endeavor in this field of Astronomy is to advance the search for the "truth" over the grand deceptions that have been taking place through the centuries and persist to this day - so help me God.

Here are some scientists with their theories that are 100% wrong, and should be discarded from all textbooks: Albert Einstein, Arthur Eddington, Isaac Newton, Stephen Hawking, and Christian Doppler. Here are scientists that are partially wrong: Edwin Hubble, Max Plank, and Schwarzschild. Here are the theories that are 100% wrong: Big Bang; Expanding Universe; Doppler Shift; Gravity; Theory of relativity; Special theory of relativity; Stars are filled with hydrogen gas; Gravity holds our atmosphere; and many more. Mathematics in Astronomy might be inapplicable. Physics in Astronomy is 100% inapplicable because 97% of the mass of the universe is Black Holes. The Physical laws that have been discovered are applicable only for "matter # 2" (atoms and molecules). The findings in the science of physics apply to "our world" and other observational phenomena, but not to the rest of the matter in the universe which is unknown to the scientists at this time. Looking at the universe they see compound particles of the atom, atoms and molecules, but not the most abundant "matter" of the universe – Basic Building Particles of the Atom (BBPA) which I also call BLACK SPHERE SUBSTANCE (BSS), which is about 97% of all matter in the universe.

SCIENTIFIC SECRETS

Can a scientist discover something about the universe, as I have, and give it to the human race? This has been a long tradition in the human history, but not without glitches. Galileo Galilei was under a house arrest because The Vatican did not like scientific knowledge not approved by the "Central Committee of the Vatican Politburo." There is a problem with the definition of "human race" which today is more like "warring human tribes" than some "unified" race. On top of it, the entire science of Astronomy consists of 85% nonsense, misconceptions, and fairy tales passed as some sort of "science" – backed by fake mathematics and fake physics. "Armed" with this kind of knowledge, today's scientists might have trouble finding where the rest rooms are. There is more to this: quasi-scientists like Einstein, Eddington, and Hubble, just to mention a few, have opened the door for many like Hawking and some of today's know-nothing "leading scientist" in the world, and the real scientists like me are considered idiots. I have a new description for a scientist: First, one should have an inquiring mind; second, one should have a "critical thinking;" and third, should be able to distinguish between reality and illusion.

SHOULD THE "BLACK HOLES" BE CALLED WITH THEIR REAL NAME "BLACK SPHERES?"

All of you have seen photographic pictures of a black hole, which is the Sun. The first layer of the Sun's atmosphere is called Photosphere which has the so-called "black spots." When we look at these black spots we are looking directly at the surface of a black hole, and it is black, but it is not a "hole?!" There is one x ray picture of the Sun from August 1973 where the edge of the surface of the Sun is clearly visible, and it is obvious that it is a SPHERE. The brain-washed scientists do not believe that the Sun could have a surface, because they are told that the stars are filled with hydrogen gas.

The guy who gave the name "black holes" had in mind the bad reputation of a prison in Calcutta where a prisoner enters alive, but does not come out alive. My objection about this name is that although it is rather poetic, it is not a scientific one. Here is an example of what is happening when things are not called with their real names: "The fabric of space around a black hole is so twisted that makes "hole" in the universe." The great idiot and mathematical crook which is accepted as a genius Einstein did not specify what the fabric of the universe is, so I assume it might be cotton. Let us call things with their real names – BLACK HOLES should be called BLACK SPHERES for one simple reason, because they are black, and they are spheres!

WE HAVE EMERGENCY IN OUR HANDS, BUT THE LEADING SCIENTISTS ARE ASLEEP AT THE WHEEL

What is known as "solid iron core" inside the Earth is its BLACK SPHERE. Unbeknown to the human race it has been "taking care" of us, and the most interesting thing is that it was taking care of us without being aware of it!

The "black sphere" (BS) inside of our planet has to be monitored 24/7 if we really want to find out what external force creates the global warming; weather havoc; and most importantly how long this might last; would it get any worse; and at the "end" would it get back to normal? Our lives depend on the proper function of this black sphere, as a matter of fact, if our BS lives and functions properly – we may live. , and uUnfortunatelyUnfortunately, there is no plan "B" for us at this moment if something wrong happens to it, and it is happening right now. At this very moment our beloved BS is in trouble, when we are not ready to become space aliens yet. I am concerned for the human's race survival, and want to give some recommendations like: Save and store some enormous amount of food to last a very long time.USE ALL LAND THAT COULD PRODUCE FOOD, AND STORE ENORMOUS AMOUNTS TO LAST 20 TO 30 YEARS! -"Earth, mother of us all!" wrote the ancient Greek play writer, and he was correct. We all have two mothers: one that gives birth and raise us, and the second one "Mother Earth" taking care of us to the rest of our lives providing shelter, food, air, and water. But according to Mayan people's calendar from 2012 we should enter in a new "era." We are in this new era right now! The simple truth is that "intruding" "influence" by "foreign celestial body" to the solar system is displacing the BS of the Earth from its rightful place, when this BSit happens to be our life-line. This displacement is making our lives miserable at this point (global warming, forest fires, etc.), and farther displacement would endanger our survival (food), and if this displacement gets quite big – we are doomed! I already have discovered the universe, but what good that would do for us, if we cannot survive in the "new world," the way we are, or instead- become space aliens?

What is theThe current location and the eventual "displacement" of the Earth's BS haves to be reported daily, because this is so important for our lives. This can be achieved through monitoring both north and south magnetic poles minute by minute, because this subject is of paramount importance for the humanour survival. What the leading scientists and their "supper computers" know about it? ... They don't have a clue!

There might be difficult times for the human race ahead of us. I neither wanted to scare you, nor do I wish to be a bearer of bad news. I simply want you to be aware of what is happening right now, and what might happen ahead, so you could be betteradequately prepared. Because at this moment we are losing atmosphere, the weather havoc probably would get worse, and the forest fire might intensify. Definitely expect new weather patterns. Crop failures are already reality, and might get worse. If I know where the "intruder" is, I can give you better predictions, but at this point I am to let you know THAT THE GLOBAL WARMING IS NOT CAUSED BY HUMAN ACTIVITIES! All this is a direct result of "intruder's" influence. The human activity makes things worse, but definitely are not the main cause, then again we should take better care of the environment.

    How long would this last, and how bad it might get depends only on how long this bad influence would continue, and how close this bad INFLUENCE ~~celestial body~~ would get to us. Some celestial bodies in the Oort Cloud are moving with the speed of 1 km per second. With this speed just the little curve around the Sun would take hundreds of years. These are wild speculations. We need to locate it; determine its speed and trajectory, and only then we can have some reliable data. The scary thing is that all the scientists in the world are not aware of what is going on, and subsequently they are ~~not looking~~ NOT LOOKING for the eventual intruder (if there is such one), and even if they see it, with their fake formulas, and being brainwashed by Newton and Einstein, they might not be able to give you any straight answers!

    The leaking of the atmosphere is very disturbing. What if it continues in the next 250 years? Already we have change in the weather patterns for the worse, and it is only reasonable to expect that with time would get even worse. As this "leaking" of atmosphere goes on, would the Earth be able to replenish these loses, when forest fires are going on, and deforestation is widespread? I have read that the stratosphere is going down from its traditional distance from the Earth, and if this is the case; we are losing atmosphere at faster rate than the environment produces. Change in the weather patterns is inevitable. If this is the case; next expectation might be that some populations living in high-altitude areas might have trouble breading as the air gets thinner and thinner. Reliable predictions could be made only if we locate the "intruder" and calculate its velocity, and only then we could know how long it would stay in the Solar System.

    To have a clear picture of the situation right now, here are some numbers: Earth's diameter 12,756 km; our black sphere diameter 1,220 km (currently mistakenly called "iron core.") The black sphere is displaced from the center of the Earth 400 to 500 km. If our BS would move farther from its base; things would get accordingly worse for the global warming, weather havoc, forest fires, and the leaking of the atmosphere. It is possible that it can get temporary back at the center of the planet where it belongs in first place and then move back out again, but what should not be happening is to get too far from the center. The situation is fluid, because the celestial body that causes this bad influence is moving. If it gets any closer, our BS might get too far from the center of the Earth, and it might reach the POINT OF NO RETURN! Just guessing, this point might be around 3,000 km from center. If we monitor both magnetic poles minute by minute; we could pin point from which direction the bad influence is coming. The Earth daily rotation also could ~~an~~ show the direction from ~~point at~~ where the intruder's influence is coming. ~~If you need answers; I need this data.~~

Usually, when we do not know something we turn to the specialists for answers, and that is how supposed to be, but what happens when the "specialists" which are the leading scientists DO NOT KNOW the subject? ~~Their knowledge in the field of Astronomy is 90% WRONG!~~ What is the decisive factor protecting our atmosphere? According to the leading scientists this is the "gravity." I totally disagree! The ultimate protector for any living atmosphere is the magnetic field! If this field is removed; the atmosphere will dissipate in due time~~,~~. Similar situation is developing right now with the Earth – in very mild form. ~~and this is happening right now, because of the BS displacement.~~ I am asking the scientists: "Mars still has its gravity, then why it lost its atmosphere?" My answer is: Because it lost its magnetic field 700,000,000 years ago when its BS was removed from the center of the planet in a similar situation. Then Mars' BS "lived" for another 670,000,000 years, but not at the center of the planet where it would have generated good magnetic field that protects planet's atmosphere, but spend the rest of its life close to the surface under the 4 biggest volcanoes in the SS in the region of "Tharsis Montes."

Let's see why the leading scientists think that atmosphere is held by gravity? Venus and Saturn's moon Titan have some sort of "nonliving" "atmospheres," and both of them do not have magnetic fields, but I will stand firm that the ultimate protection of any living atmosphere is well-working magnetic field! BS displaced from the center of a planet cannot produce quality magnetic field. All planets have gravitation, regardless do they have working BS inside or not. There are requirements for good magnetic field to function. The first one is that the planet should rotate. One of my universal laws states: "The mantle together with the iron around the black spheres of a planets in any "'Solar System'" rotate~~is made to rotate~~ around their axes, because ~~by~~ the 'Sun' makes them do it." Another of my laws states: "The black spheres inside all celestial bodies are ordered not to rotate around their axes by the Sun." Thanks to these "orders" and their enforcement we have good magnetic field that protects our atmosphere. But here comes an intruder with its influence and "orders" that are different than the Sun's orders, and that is how the planets in the Solar System currently are subjected to contradictory influences of two powerful bodies. ~~700,000,000 years ago in similar circumstances Mars lost its atmosphere.~~ The bad influence of intruding celestial body 700 million years ago ~~kill~~kills Mars' atmosphere. At this moment ~~T~~the same story in smaller scale ~~at this moment~~ is happening to the Earth. Let us keep our fingers crossed that our BS is not yanked out of the center of the Earth any farther.

Let us look first at the known facts, and then do the speculations. Our Solar System is under "EXTERNAL INFLUENCE." All planets, brown dwarfs, moons, asteroids, and comets are NEGATIVELY affected – including the Sun! What is this thing, and where is it? At this moment we can only speculate. Most likely this is a big brown dwarf which at this moment is inside the Solar System which allegedly is 15 trillion km across, and if this thing is so big, how come that with all these telescopes and satellites we cannot see it? Caution; do not bet all your money on the existence of this "intruder," because this bad influence could come from outside o~~i~~f the solar system. From outside the SS the negative influence cannot come from planets or brown dwarfs – they are too weak. The possible players from outside the SS are white dwarfs, stars, and neutron stars. Black spheres could affect the SS even if they are located outside of the Galaxy. ‑The "intruder" might be a "brown dwarf," and we have 5 or 6 of them right here in the Solar System – the four gas giants plus the Jupiter's moon Io. The strength of a brown dwarf is directly proportional to the size of its "core" – the bigger the stronger. Jupiter's "core" reportedly is large but the data is insufficient. Allegedly the entire solid body of Jupiter is about 10 times the size of the Earth, and the outer size is ~~looks~~ big, but actually it is puffed-up thing – it is basically gas. ~~But let us not forget the "white dwarfs" which are 1,000 times more powerful compared to their size.~~

Here is one curious "fact~~:~~." There ~~should be two~~could be many brown dwarfs with the same size like Jupiter~~,~~; ~~one should be weak and~~ and the ~~other one~~strongest of them could be 130 times stronger than the weakest. ~~Of course there would be many "hybrids" in between.~~ Here is another curious fact: The weaker one should have way stronger magnetic field, and this would cause way bigger atmosphere protected by it.

The fact that the two peripheral "planets" Uranus and Neptune have been affected earlier than us means that the "intruder" was closer to them, and the fact that now the Earth is getting more and more affected means that the "intruder" is coming closer to us, and that could give some clue where this "thing" might be right now.‑ If the intruder affects the most powerful celestial bodies in the SS means that is way bigger than them. By the way, all gas-giants are way stronger than the Earth, but their "strength" is insignificant in the big picture. This intruder is affecting every celestial body in the Solar System (SS). Looks like such strong external influences whether from outside or inside the SS ~~intruders~~ have happened~~been here~~ before, and upon "leaving" the SS, they "leave" planets like Uranus' with one of its magnetic poles "turned" toward the Sun. The scientists say that this is caused by some sort of collisions, but I am reporting to you that powerful enough influence from outside celestial body ~~of the~~ SS causes this phenomenon through challenging the established order of the Sun ~~in the SS~~. Same is true about the 23 degrees incline of the Earth ax~~i~~es of rotation.

We have to locate it, and observe it. This would give us important data on what its speed is, how long it will stay in the Solar System, and the most importantly how long would affect negatively our planet. Keep in mind that the negative influence on~~to~~ the Earth is directly proportional to how close this thing would get to us.

Let us look at second possibility. First of ~~all~~all, what "affects" the celestial bodies in the Solar System is another black sphere, because all celestial bodies in the universe are eider black spheres (could be disguised), or derive their existence from it. I am 100% sure about that. The only question is where it is and how big it is. This "thing" affects the Sun as well, and because the Sun supposedly should be affected "differently" than the other "members" of the Solar System would give us a clue about its size. If the Sun is not that much affected, then it should be a brown dwarf or its~~the~~ equivalent of white dwarf. But if the Sun is affected too much, then this could be enormous brown dwarf with BS 300,000 km diameter and up. Interestingly, its equivalent of "whit dwarf" should be some 300 km. ~~or a star outside of the SS. This influence could come from big BS which might be located at the edge of our Galaxy, but I doubt that, because of the speedy movement of the North magnetic pole.~~

PREPARATIONS TO BECOME SPACE ALIENS

As weird as it might ~~sounds~~sound, sooner or later, we have to begin preparations to become space aliens. Initially, I would advice for a better preparedness for a survival – storing food for 30 years or so. The Earth cannot be always hospitable to us. It might die suddenly, and the Sun as well.~~;~~ In the current situation ~~This death could be final and permanent, or could be~~ we might be temporarily deprived of food or shelter. ~~temporary unable to provide us with food and secure shelter.~~ The permanent death could come from the death of these two very important BSs in our lives – the Earth's and the Sun's. Unwelcome and strong influence is threatening the SS established order right now. Keep in mind that the life of our Galaxy and for that matter the life of the entire local group of galaxies is scheduled to come to an end in the next 500 million ~~1 billion~~ years or so, and the "local crunch" is going on right now. This "end" does not come with some absolute and exact precision; for example, some celestial bodies like the Moon and Mars are already dead. I would speculate that the Earth and the Sun might or might not live another 500 million years, but the END is knocking at the door. An intruding influence of sizable BS can destroy the established order of the SS. My suspicion is that civilizations that become space aliens did that out of necessity rather than any other reason. A civilization has to be technically sufficiently advanced and prepared to become "space-alien civilization." I am trying to asses how ready the human race is in order to become space alien race. We are not ready at all. We cannot become space aliens at this time. ~~Firstly~~Firstly, we do not have the transportation. Yes, we do have rockets, and obviously we can go to the space station and the Moon, but beyond that we are having trouble. No space aliens have arrived to the Earth on rockets yet (this is a dilapidated technology) – maybe they had rockets as well but millions of years ago. Development of UFOs should be our ultimate transportation goal, but the human race is not united.

Let's examine our morality: DOGS ARE WAY MORE MORAL THAN US. We are liars, cheaters, back-stabbers, morally bankrupt, murders, thieves, racists, potential cannibals, deriving pleasure from torture other human beings, dumb and evil, unable to govern ourselves, collectively dumb, and no democracy or dictatorship can help us. We are deliberate liars, careerists, fascists, back-stabbers, secretly hating each other to death. Words like "consciousness" and "common sense" are total misnomers. We do not have a moral compass, and in a split second we are ready to become fascist criminals.

Our planet is a jewel, but now is in trouble. The reality is that our planet is very fragile jewel, and that makes our lives uncertain. Although the Earth is the biggest "planet" in the entire universe; compared to other celestial bodies is very small and FRAGILE. From the dangers we are surrounded let's take a closer look at these supernovas. I am reading a textbook that says "unfortunately there are no supernovas close by so we can study them." I would say: How fortunate we are that no such a monster has exploded near us. Funny how the writers of these books always see some beauty in these horrendous explosions. After a supernova explosion many planetary systems around it are completely destroyed. The planet's "mantles" are thrown into space. Funny thing is that the scientists see some "star formation" where there is only total planetary devastation. I want to know how many light years of devastation are done around one of these monsters called "supernovas?" I am not sure, but it seems to me that the space aliens prefer to live underground or on the bottom of the ocean, and assuming that they have been around way longer as advanced civilizations, they do not like to be exposed to dangers. Are we endangering our lives living only under the "sheltering sky?"

Let's say that it is high time that we begin preparations to become space aliens - nothing wrong to be ready just in case. Our first destination should be our Moon. If we can make it in the Moon, we could make it on other similar celestial bodies. Several points as to why first we have to "conquer" colonize the Moon. Here are some questions: Can children be born and raised there? Can people living on the Moon raise their own food and not expect deliveries from the Earth, which in case of emergency might not be available? We do not have the vehicles to go any farther than the Moon, but in this book you will learn that the energy in the universe is free. All we have to do is tap into it. On other celestial bodies like the moons of the gas giants where the Sun cannot provide enough heat; the necessary energy for survival is available only from the vents of volcanoes, when the surface temperature of this celestial body might be minus 168 degrees C. Our Moon does not have them, so this valuable skill for the future space travelers has to be learned here on Earth. Once we learn this skill; we can go in space explorations because almost all small celestial bodies which are not bigger than the Earth could provide the necessary reliable energy as long as they have active volcanoes which are indication of live and working BS inside. Through knowing what chemical elements and compounds are coming from them, the space travelers can count on their reliable supplies.

THE GODS WISH YOU WELL

The war between good and evil never ends. Let us say that knowledge is good, and deception is evil. The disagreements I have with the scientific community in this field are too profound and irreconcilable. God wants the human race to prevail. The universe is our home, and the knowledge about it has to be correct. In this book I would vilify some people like Einstein and Hawking not for any other reason, but the fact that they did not know anything about the universe, yet decided to explain it to you, and in this process mislead the human race through spreading lies and misconceptions. Finally, it is all in the hands of God, and he wants the human race to have the true knowledge about the universe, and I consider myself a servant of God. I have to deliver the "scientific truth" to you regardless that humans are treating me like a disposable diaper, and I cannot be but bitter about it. The scientists in this field are severely brain-washed (85%). The situation is urgent and calls for an outsider, so here I come to the rescue! This book cannot clean this enormous swamp of "scientific" misconceptions. So this might be just the beginning. The Renaissance in Astronomy might start with this book.

THE UNIFYING THEORY OF EVERYTHING HAS BEEN DISCOVERED

From Wikipedia: "Over the past two centuries two theoretical frameworks have been developed that together most closely resemble a theory of everything." And these are the "general relativity" and "quantum mechanics." General relativity focuses only on gravity, and quantum mechanics on non-gravitational forces subatomic particles, atoms, and molecules. Another quote: "General relativity and quantum mechanics have been repeatedly validated in their separate fields of relevance." In pursuit of getting these separate theories together, some sort of "quantum gravity" has been researched, but NEVER FOUND!

Here is my take. Quantum mechanics is real, but is applicable only to 3% of the matter in the universe. Newton's gravity is 100% baloney! Mentally deficient Einstein "improved" it, and Hawking with mental ability of a six-year old broadcasted it, and to the existing crap he added some of his own! Wikipedia states that "general relativity" was continuously "validated." I wonder how they "validate" complete and total nonsense!

The universe is in the state of "perpetual motion;" forever "alive" repeating one and the same "universal cycle." Reason for this might be the PERPETUAL IMBALANCE of the Basic Building Particles of the Atom (BBPA). We might be able to find some control over the entire universe, and it could be some perpetual live imbalance of the basic building particles of the atom BBPA. The "unifying" of the entire universe lies in its "uniformity," but for the galaxies to get together in order to "complete" the Grand Universal Cycle there is a local "central command." This "command" is either on the super cluster level or beyond, but it is "local" and

~~identical to myriads similar once.~~ Yes, there can be things that unify the entire universe: There might be a uniform magnetic field equally strong throughout the entire universe. The notions of parallel universes and universe being spherical are unrealistic. The universe is one giant system with many smaller systems encompassing each other.

Philosophically speaking I want to point out why the scientists cannot see "the reality." Whatever they see, they "pass" it through the "prism" of their nonsensical Big Bang; then through the prism of 100% no existing "gravity;" then through the "expanding universe" NONSENSE, when the universe is expanding only in the heads of some people but not in reality; then they look through the prism of mathematics (Who are we to impose our mathematical rules on the universe?); then comes the prism of the absolute nonsense of "stars are filled with hydrogen gas;" then comes the prism of the claim that "the entire universe is made of atoms and molecules." I do not care how smart a person can be; if he accepts all this nonsense as true statements, there is no way in the world that he can see or discover anything, or think straight for that matter! But that is what the textbooks in universities preach, and the brain-washed professors in the most prestigious universities all over the world brain-wash the "paying" students.

THE TRUE UNIVERSE DECIPHERED

Welcome to the real universe, and not the one "invented" by some smarter than monkeys "Einsteinses" and "Hawkinoids." Is there a live BLACK SPHERE inside the Earth? Yes, there is, right at its center. The scientists are calling it "solid iron core" as they do not know what ~~it is~~ it? The diameter is 1,220 km. At this moment we are alive because our black sphere is functioning, and if it stop working~~dies~~ or something happens to it – we would ~~might~~ die as well, or become space aliens if we can. Without live BS inside - the Earth would look like Mars. Mars lost its BS, and that is the reason why it lost its living atmosphere and surface water.

Is there a BS inside our Moon? Not at this moment – it has done its job of creating the Moon, so its mass now is the mass of the Moon. Is the Sun a black sphere? Yes, it is, because it is a star, and all stars are 100% black spheres. The current belief that all stars are full of hydrogen gas is absolute and complete nonsense! Each planet and moon was created and begun its existence with a BS inside it. There is NO ANY OTHER WAY of creating planets and moons, but with black spheres present inside them at the moment of their creation, ~~WHEN A GALAXY IS CEATED~~. Today's scientists believe that some rocks get together forming so called "planeticimals," which get together and form planets. They also erroneously believe that stars are created from hydrogen gas at all times. These are the fairy tales of ignorance! Stars are created only at the galaxies initial creation; basically they are black spheres.

Basic Building Particles of the Atom (BBPA) lives forever, and creates everything that we see, which are all atoms, planets, moons, and humans among other things. All these things that surround us are "temporary creations," and one day must to be "destroyed," because this is in the "script" of the Grand Universal Design (GUD). This sounds weird, but it is true: whatever the black spheres create, they destroy it through breaking each and every atom to basic building particles, and stashing them back into the black spheres from where they came in first place. They allow the existence of atoms and molecules for around 5 billion years before destroying them completely. Then, believe it or not, the black holes are creating atoms, planets, and moons again. So, they "destroyed" them, then create the same stuff again, isn't that stupid? Call it any way you like, but that's what is happening. All this is part of the Grand Universal Cycle (GUC) lasting about 20 billion years. The constant repetition of this cycle in different clusters of galaxies at different times defines the life of the entire universe. Shouldn't the black spheres run out of energy doing all this? Are you ready for a shocker? THE ENERGY IN THE ENTIRE UNIVERSE IS FREE FOREVER. There is no such thing as matter becoming energy and vice versa, and Einstein's fames formula ($e = mc2$) should go to its rightful place - the trash. By the way, I talk to one guy working in the universe's maintenance department. I asked him, why this "universal cycle" has to be repeated constantly, and he answered: "I don't know - I just work here."

To understand these new discoveries, one has to be open-minded because they are quite contrary to what the scientists have made us believe, then again, you remember that the Earth at some time was "flat"; then it was the "center" of the universe; then the "center" of the Solar System?

Tomorrow humans would wonder could there be dumber idea than the Big Bang? It is time for another leap forward in our understanding of the universe, but we have a tendency to reject the new and controversial discoveries almost instantaneously, especially if it is coming from outsider like me. Keep in mind that those who studied Astronomy, and today are professors, having PHD, writing books, and are on TV, have been brainwashed right at their initial education, and today they are brainwashing the young generation in their turn.

The job for "scientific investigation" asks for a person outside of the scientific "mafia." Was S. Hawking from this mafia? He was "theoretical physicist" so was his mentor Einstein. I wonder what qualifications are needed to become a "theoretical physicist" except being an idiot who is "entitled" to "explain" the universe to us, although he doesn't know anything about it! According to Hawking: "From the dawn of civilization, people have asked questions like: Did the universe have a beginning in time?" This is a good question for "nonsense-talkers" like Einstein and Hawking. I doubt that people asked such impractical questions, but sure enough such questions are very useful for "writing books" and being "famous" for discovering nothing. What these scientists are doing is talking to themselves, when in the same time are deeply sunk in their own stupidity. They are giving explanations about the universe as if they are the "smart," and we are the "stupid." Thank you, Isaac Newton, Einstein, Eddington, Hawkings, Doppler, E. Hubble, thank you for all the nonsense you have said about the universe.

I was in a bookstore in Minsk, Belarus, and saw abut 8 books from S. Hawkings. I do not doubt his intelligence to tie-up his shoes, but he is one true-believer who does not question any of his predecessors. He said that he is standing on the "shoulders of giants," and one of them is the mentally-deficient-brain-damaged mathematical crook Einstein. The other "giant" is Isaac Newton, who imagined that he discovered the "gravity of the universe." As far as Earth's gravity; any dog knows that if something is thrown up; eventually will go down. The gravity is part of the overall "influence" of a celestial body, and it does not extend farther from the Roche's limit.

I think that I HAVE DISCOVERED THE TRUE NATURE OF THE UNIVERSE and it is boringly simple, but there are many details that still have to be discovered. As the new model of the universe is to be affirmed, many of the old beliefs have to be disposed of, and they are quite a lot. I am reading from a current textbook for advanced college students, and I see that almost all their MATHEMATICS AND PHYSICS ARE WRONG!!! This is happening because of many wrong assumptions through the centuries that have been taken as "reality."

Here is a partial list of the biggest spreaders of scientific lies and misconceptions: S. Hawking, Einstein, Eddington, Hubble, Doppler, and Newton. All of them being 100% WRONG. Now let us look into the sewer of some of the biggest and most deceptive theories considered "reality" at this moment: Newton's gravity; Big Bang; expanding of the universe; Doppler shift; stars are powered by hydrogen gas turning to helium; stars are born in hydrogen clouds right now; planets are formed from rocks getting together; and the Earth's atmosphere is held by the gravity. All this is ABSOLUTE BALONEY!

Newton's "gravity" is invalid outside of the field of falling objects to the ground. The force that is pulling things to the surface of the Earth is not the same force that keeps the planets going around the Sun, or the Moon going around the Earth. I have discovered the true "influence" of the celestial bodies in the universe which I call a "UNIVERSAL COMMANDING INFLUENCE."

THE UNIVERSE HAS EXISTED FOREVER, AND WILL EXISTS FOREVER. IT IS TIME TO PUT THE BIG BANG TO REST.

There was no "Big Bang" as they claim; instead there was one of the many BIG UNIVERSAL AND LOCAL EXPLOSIONS. The universe is stationary as a whole, but it does go through constant changes, when in the big picture nothing substantially changes ever; just different clusters of galaxies go through one and the same "cycle" of events at different times. Let's say that we can go through time, and let's say that we are going back in time 50 billion years. We land on a planet like the Earth, and look at the sky. We see almost the "same" stars and galaxies. Someone asks: "Where is the Milky Way?" The answer is that the Milky Way has not been created yet. Let's go now 50 billion years into the future, and the question now is: "Where are Andromeda and the Milky Way?" The answer is that Andromeda and the Milky Way no longer exist, because they have been recycled, and new galaxies were created in their place. The universe goes through constant repetition of the same UNIVERSAL CYCLE, and the duration of it might be around 20 billion years. Basically it consists of constant recycling of the galaxies and everything inside them. Why this is happening is yet to be discovered.

There are three kinds of matter in the universe. One is the matter we all know which consists of atoms and molecules, which consists of the iron around the "core," the mantle, the crust, and the atmosphere. This is matter # 2. The other matter is "matter # 1," and is in all stars and the centers of all "live" planets and moons. The third matter is "compacted matter #1." This is the densest possible matter in the Universe which is found in neutron stars and white dwarfs (matter # 3). The planets and moons that are "alive" have two different matters in it: The matter in the core which is BLACK SPHERE SUBSTANCE (BSS), and the matter in the mantle and the crust made of atoms and molecules. Because the human race does not have a clear understanding of BSS and the existence of black spheres inside all "live" celestial bodies, they erroneously assume that ALL CELESTIAL BODIES are made only from atoms and molecules, but that is not the case. So the presumption that what we discover in the laboratories is the same "stuff" that the entire universe is made of is 100% WRONG. Now all mathematical calculations and physical laws have to be reevaluated or outright discarded for this simple reason. The total volume in the universe of matter # 2 is between 0.5% and 3%. The matter # 1 in the universe which by volume is about 97% is yet unknown in detail to the human race. In these 97% is included matter # 3 which percentage is to be determined.

My impression is that this constant repetition of this UNIVERSAL CYCLE is dictated by some sort of "necessity" of "matter # 1" to be in the constant state of change and perpetual "dynamic existence." So "matter # 1" is permanently "unstable" and requires this constant "reorganization" which gives the appearance of something that is "alive," which never come to a state of equilibrium. In this case we can say that the universe is "alive" and "lives" forever! It is going through many GRAND UNIVERSAL CYCLES at different locations in the universe (clusters of galaxies). In these constantly repeating this cycle "matter # 1" goes through constant "transformations" and "transfigurations" which we see; they are always the same, but some confusion in our conclusions is possible. The first confusion comes from the change of its

18

density, because of the "matter number one," as "matter # 1" CHANGES its density because it constantly releases BASIC BILDING PARTICLES OF THE ATOM (BBPA), and in this process constantly looses mass. The released BBPA are creating compound building particles, atoms, and molecules called which I named "matter # 2." The creation of matter # 2 is nuclear in nature, and scientists do not know anything about it. The human race exists in it, knows and studies it, but it is only "matter number two," and erroneously assumes that the entire universe is made only from this matter, but that is not the case!

Through repeating the same cycle in different locations forever, the UNIVERSE LIVES FOREVER! That brings the first question: Shouldn't the universe one day run out of energy? If this has not happened yet, then there is only one possible conclusion: The energy in the entire universe is ABSOLUTELY FREE, AND MATTER AND ENERGY ARE INDESTRUCTABLE AND INSEPARABLE!

"Matter # 2" is created temporarily, and it is constantly created; then destroyed; then created again and destroyed again – f forever; with several billion years in between. That brings the question: What lives "forever?" THE BASIC BUILDING PARTICLES OF THE ATOM LIVE FOREVER! The universe never loses the most insignificant amount of energy or matter ever! Then what is happening to the conversion of matter to energy and vice versa? Could it be that "e = mc2" is BALONEY? Yes, it is! We have to distinguish between BBPA and the word "matter," because the word "matter" also implies atoms and molecules, and if we say: "Matter and energy are indestructible," then this statement would be inaccurate, because atoms are "destructible." For example, the defenders of the "famous" formula are claiming that the mass of two hydrogen atoms is more than the mass of one helium atom that is produced from them, and that is their a "proof" of its validity. I say, that this is not the case: The matter that supposed to be missing is particles which through their "separation" have released this "atomic" energy, but in this process there is no "conversion" of matter to energy. The missing particles from this reaction have gone somewhere else. Where they might be? The neutrinos which allegedly have some mass have arrived in 1987 from 168,000 light years from the site of a supernova explosion (Large Magellanic Cloud). So if you want to account for the missing mass from this explosion, you have to account for mass flying hundreds of thousand light years away from the explosion in each and every direction. How many neutrinos are flying away from this explosion? How big is this "ball" with radius if 168,000 light years? Then to account for all the neutrinos one has to look for them in the diameter of 336,000 light years! The Earth is really small, yet many neutrinos are hitting some small drums. How many neutrinos are released in this supernova explosion, and allegedly they have mass?!

The smallest particles of the atom which are BBPA plus their energy are the two sides of the same coin. The particles are indestructible, and the energy somehow is always with them. The universe runs its operations without the loss of any energy. The key to understand this enigma lies in the black sphere structure.

The universe is one giant well-organized system, where almost nothing happens by chance. Probably the universe is commanded by some decentralized principle which is based on electromagnetism, energy, and matter. The universe is runaway special nuclear reaction creating the illusion that is "alive," but it is in this perpetual motion forever.

All the stars are black spheres, therefore all of them are filled with the same BLACK SPHERE SUBSTANCE (BSS), and the BSS consists of BASIC BUILDING PARTICLES OF THE ATOM (BBPA) which are arranged in some special way which is a mystery at this time. BBPA are not stashed like potatoes in a sack, because the black spheres perform different tasks at different times – therefore BSs are some sort of a "machines." Here are some tasks that BSS performs: Releasing BBPA at different rates, where the rule is – the bigger the size of the black sphere; the ~~faster the~~ faster the ~~rare of~~ release ~~is~~. This rule dictates what celestial body would be created, and what role it would play in the Grand Universal Cycle based on its size only. The smallest BSs would create small spherical stones; then bigger BSs would create moons and planets; after that come the brown dwarfs, which begin the process of rejecting the accumulation of "matter # 2" on their surfaces until comes the total rejection, and these are the smallest stars.

Except afore mentioned "normal" black spheres; there are "compacted" black spheres, which are "produced" after a supernova explosions and maybe at the initial creation of the first galaxies after the initial explosion of the biggest black spheres. Their behavior is totally different. They act as if they are "thirsty and hungry" and suck up material from other stars. Maybe through this process they become normal black spheres, thus from "matter #3" (compacted) become "matter # 2" (regular BSS)

The second thing that the BSS does is "dismantling" of any whole atoms to its BBPA upon entering in a BS. We could learn about this process only through observations, because no lab experiments are applicable, and no mathematical or physical law should be applied, as they are invalid! Computes and "artificial intelligence" are strictly prohibited! Nonsense in; nonsense out! These jets of gamma rays that are seen coming from the both polls of a star and central black spheres are some sort of particles or some unnecessary energy that has to be taken out at this time, but as the whole universe is one big system; ~~every particle~~ in a long run every particle is accounted for.

Energy is omnipresent and never lost! Here is the Sun creating a "wind" of BBPA, and doing this for the last 13 billion years or more. Through all this time enormous amount of energy has been "released." This energy is created from the process of creating "matter # 2" from BBPA. Humans do not have a clear perception of this process at this time. For them the energy coming out of the stars is derived from matter, which converts to energy, and this energy is lost forever. What is the reality? Firstly, matter never converts to energy; and second energy is never "lost." Energy is simply ever present. ~~Now the exact details of how energy is never lost, have to be worked out, but unquestionably, energy is ever present and never lost in the most insignificant amount, otherwise the universe would have ceased to exist by now~~!

What is clear is that energy is created firstly when the BBPA are self-assembling once pushed outside of the stars, but when atoms are disassembled at the end of the cycle again energy is released, so the energy is forever present regardless whether atoms and molecules are created or dismantled. Why the scientists think otherwise? They "live" in the world known to them; in order to create one chemical element from another in many cases energy from outside is necessary. In reality it is not only unnecessary, but when the BBPA self-assemble in atoms external energy is "created;" then in the reverse process atoms are dismantled back to BBPA; ~~and a~~again extra energy is "released." Both of these processes are unknown and not investigated fully at this moment. And that is how the energy is ever-present and ever-lasting. We have a faulty preconditioned "understanding" of "energy."

We have galaxies that lack any matter # 2 (atoms and molecules). When and how all atoms were dismantled, remains an open question. The world we live-in which is the Earth and everything on top of it is "scheduled" one day to "disappear" and "recycled."

Let us say that a GRAND UNIVERSAL DESIGNE exists and part of it ~~are~~is all these~~many~~ LOCAL UNINERSAL CYCLES which are absolutely the same but separate for each cluster of galaxies. Each of UNIVERSAL CYCLES begins with its separate BIG UNIVERSAL AND LOCAL EXPLOSION that creates galaxies. All of these explosions are the same.

It had been estimated that the Solar System began its existence 4.6 billion years ago, and the Galaxy began its existence 13.8 billion years ago. There is~~but I have~~ evidence that both of them began their existence TOGETHER; therefore the Solar System began its life 13.8 billion years ago! ~~it must have began its existence with the initial formation of our Galaxy, and this gap of 9 billion years has to be closed somehow.~~ Here is a new~~So here is my~~ hypothesis: After the BIG UNIVERSAL EXPLOSION, our Galaxy was "lentil," and existed as such for 9 billion years. Immediately after the explosion all black spheres are "dysfunctional" (not creating matter # 2) because they were "compacted" by the biggest universal and local explosion (BULE) and needed time to "recover" – matter # 3 to become matter # 1 2. Our Sun was 1,000 times smaller than is right now, which mean as big as the Earth. The Earth was 1.2 kilometer (the Earth's BS is 1,220 km)divided by 1,000). At that time matter # 2 was not created yet. What kind our Galaxy initially was, I am not sure, but was not "spiral." ~~definitely did not look like right now. What I am sure i~~For sure it was~~s that was~~ the same size, because although the Black

Spheres were compacted (matter # 3), THE INFLUENCE OF THE BBPA NEVER CHANGES. That means that a celestial body might change its size, but the INFLUENCE OF ITS BBPA REMAINS THE SAME.

The Earth did not have any mantle, which is matter # 2 of stones and metals. They could be produced only by "functional" Black Spheres, which cannot be matter # 3, but only matter # 1. The Sun and the Earth was somewhat "dysfunctional." The Sun was not emitting any heat, because was not releasing any BBPA, as a white dwarf at that time, but most definitely had the same commanding influence over the entire solar system (SS). After several billion years our Galaxy become "spiral," and at that time the Solar System appeared as it is right now. 4.6 billion years ago the Earth and the Moon were already created but were smaller in size than right now, but the Sun's size did not change since then. The size of the 4 "gas giants" did not increase too much as well. During the first 9 billion years before our Galaxy to become "spiral" substantial amount of "matter # 2" was created on top of all rocky planets and moons; in a sense 4.6 billion years ago the size of the entire planet Earth was smaller than right now, but its BS was not "naked."

Our Galaxy would exists until it collides with Andromeda and get together with all the 40 galaxies in the Local Group to become part of one huge elliptical galaxy which on its turn would be a part of a quasar. This quasar will become one of the biggest black spheres in the universe; ready for the next Big Universal and Local Explosion that would begin the next Cycle. We could say that we are in the early stages of the local "crunch" when our Galaxy together with the 39 in the Local Group will form this enormous elliptical galaxy. Slowly no atoms, molecules, planets etc. are allowed to exist. Thus all planets, molecules, and atoms would be totally disassembled to BBPA, and stashed back into the black spheres from where they came in first place. In the quasars probably some final purification is done to the Black Sphere Substance. And this particular Universal Cycle is repeated to perpetuity. In this process no mass of a single particle is ever lost, and that is how the matter in the universe is indestructible. After "quasar stage" follows the creation of one of the biggest black spheres in the universe which I gave the name "THE BIG BLACK SPHERE" (BBS).

BBS would explode and create approximately 40 new galaxies. These explosions are the biggest in the universe called BIG UNIVERSAL AND LOCAL EXPLOSION (BULE). These big explosions occur throughout the universe every 20 billion years for different clusters and different locations at different times. Can this be connected to the background microwave radiation? Could the big explosion from the local BULE 13.8 billion years ago might have been mistaken as some sort of a "Big Bang" nonsense that "created the entire universe?!" No explosion as big as it might be could create the ENTIRE universe, which size is unknown. Here is some of the nonsense you are going to hear from the defenders of the "big bang."

Does this sound believable to you? "Out of nothing came everything." The "nothing" is as big as a walnut, and the "everything" is trillions upon trillions of galaxies and stars. Only complete idiots can believe such nonsense. In my theory all particles of the atom are accounted for, and space and time have continuous existence. Here is another one: Before the big bang there was "no space and time," may be there were no idiots like Einstein as well. We are probably the laughing stock of the space aliens. A lot of ink has been wasted on what happened in the first second after the big bang. The answer is nothing has happened, because there was no big bang in the first place.

Expanding universe was next "necessary" nonsense in order to "confirm" the "Big Bang." In 1842 Christian Doppler, a mathematical professor in Prague university expressed the idea that if a planet or a galaxy is coming toward you the light coming from it would acquire blue shift, and if it is going away from you it would acquire red shift. This is absolute nonsense; but I do not blame Mr. Doppler that much, he had an idea and expressed it. It happened that his idea was supporting the official party-line of "expanding universe," which on its part gives credence to the "Big Bang," and was adopted. It is a fact that light coming from remote sources acquires more red shift, but this is happening because the light is going through more dust, gas, and particles, and has nothing to do with the direction of the movement of the stars or galaxies. But they labeled this phenomenon "Doppler shift," and that is how it got this FIXED INTERPRETATION: When there is red shift, a celestial body is moving away from an observer, and if it has blue shift; it is coming toward him. Let us put this statement through a logical test. The speed of light is 300,000 kilometers per second – in one second and one third of it a light goes from the Earth to the Moon – quite a speed, don't you agree? With what speed a celestial body has to move away from you in order to acquire red shift? Let us look at the speed of some celestial bodies. The Sun is moving with 220 km/s going around the central BS - could this speed cause the change in the spectrum of the light coming from it, I would say NO! Second, the light consists of separate colors, and each color has its own speed as well, but these speeds are real close. As a matter of fact I would suggest to the scientists an experiment – create an equipment to register the first color of light coming from very remote source, and you might see that the blue color would come before the red. Each color is separate and the notion that they can switch from one to another – is basically stupid. Each color has its own speed, but the scientists think that all colors move with the same speed, and Einstein explicitly stated that. Then they coined the expression: "The farther they are, the faster they move away from us." They base this assumption on the fact that: "The farther they are the more red-shift they would acquire." Then they come to the inevitable conclusion that galaxies that are so far from us that we even do not see; would have to move away from us with speeds HIGHER THAN THE SPEED OF LIGHT – you see – the stupidity of this idea begins to catch-up with them. But the scientific community is not about to fall on its back, and is patching up this nonsense with another one which is called "dark energy." And guess what this dark energy does; allegedly it does not affect galaxies, but only the space between them – how idiotically convenient! All the nonsense from Big Bang, expanding universe, Doppler shifts, E. Hubble's "law" of expansion of the universe, and dark matter and dark energy are nothing else but nonsense upon nonsense! All college professors, hosts of scientific shows on TV, book writers, PHD holders, have something to "chew" on. E. Hubble invented a "constant" for expanding of the universe, but they call it a "law" – the humans make the laws of the universe, and the universe does not even know about it. The young people in colleges have something to study as well. This book has to lay the foundations of "scientific truth," otherwise the spreaders of nonsense and scientific lies will go on lying.

Let's finish with the erroneous interpretation of so-called "Doppler Shift." We have an erroneous interpretation of naturally occurring phenomena. The farther some galaxies are from us, the more redredder shift their light acquires because through more dust and particles they must go. First of all, have you noticed that in sunset the sky gets red, and why this is happening? Because the light has to go through more dust, and as a result of it, acquires this red hue. The same thing happens when there is a huge forest fire. Wouldn't be logical that light coming from the most remote areas of the universe would go through more particles, gas, etc, and get more red-shifted? Of course, so remember, the ONLY REASON WE HAVE RED OR BLUE SHIFT IS THE AMOUNT OF DUST THE LIGHT HAS TO GET THROUGH, and has nothing to do with this "celestial bodies moving toward you or moving away from you." What the great book writer S. Hawking think about it? "The only reasonable explanation of this was that the galaxies are moving away from us." It is reasonable to reason, if someone can reason, but no one told Mr. Hawking that he cannotis a moron. One might object to my language, but HeMr. Hawking has to be judged from the prospective that he was teaching the world what the universe is when he knew next to nothing about it. He was widely accepted and honored through the "scientific" world, but let me render my opinion: He is a superficial thinker; his reasoning is shallow; he lacks the ability to have a real grasp and understanding of the matter discussed. To me his conclusions are predictable. He read, memorized, and rehashed, but is far from THINKING AND DISCOVERING!

Let us go back to the Doppler Shift nonsense, and the "explanations" offered for this phenomenon. Scientists observe two celestial bodies where one of them orbits the otherrotating around each other;; and let us say that the observer's sight happens to be in the plane of the rotation. When the rotating body's light gets from more polluted area to less polluted, moving from the back of the other body toward the observer, blue shift is observed. This blue shift has nothing to do with the body coming toward the observer, but has everything to do with the fact that is moving from more polluted area to less polluted one. Then the same celestial body which

24

has come in front of the other body happens to be between the observer and the other body. Now it begins to go around from the front to the back; thus from less polluted area into more polluted aria and gets red-shifted. Any interpretation that the rotating body gets blue shift because is coming toward you, or gets some red shift because is moving away from you is absolute and complete nonsense, but the scientists believe it.

Let's now look at the other "culprit" Edwin Hubble. There is a "Hubble law" of the expansion of the universe, and by now you know my opinion about this "expanding nonsense." No such thing is happening whatsoever. The textbook is saying: "He proved this expansion to us." He has proved it to you, but he has not "proved" anything to me, because I am currently in the role of a scientific inspector. Real discoveries in the field of Astronomy may occur only if the nonsensical ideas are discarded once and for all. Hubble was looking at galaxies and noticed different red shifts coming from the surrounding gases.

On another subject, the established scientists are giving you overly-rosy scenarios for our future due to their ignorance. Here are some examples: They do not understand what role the Earth's magnetic field plays for protecting our living atmosphere, as a result of it they give the projection that our atmosphere would be around for another 1 billion years – complete baloney. Currently they believe erroneously that the gravity is the decisive factor that holds our atmosphere, when the decisive factor is the electromagnetic field. Another unrealistically rosy scenario you are hearing from them is that the Sun will shine for another 5 to 6 billion year. Reason for that is that they do not know what is inside of the Sun; and it is not hydrogen gas as they believe, but specially arranged basic building particles of the atom, or as I call it black sphere substance (BSS). It exists forever, but in small stars, brown dwarfs, and especially planets and moons ostensibly "dies" or "disappears" when actually is turning to "matter # 2" which is our world of atoms and molecules. The scientists do not have the foggiest idea what is inside the Sun, and how long would it last!

Another thing that they do not understand is the fait of all galaxies and stars throughout the universe. They assume that one day all stars and galaxies will burn out, and the sky would get dark, when actually all stars and galaxies are constantly going through the GRAND UNIVERSAL CYCLE. All planets, stars, and galaxies go through this never-ending CYCLE of "apparent" "self-creation" and "self-destruction."

A JOKE ABOUT THE SHAPE OF THE UNIVERSE AS A SPHERE

Thanks to the moronic wisdom of the idiot-considered-genius Einstein, the shape of the universe is spherical, and if you fly in a rocket in a straight line into space eventually you will arrive at the same spot from where you have started. But GUESS WHAT? One guy succeeded in developing a binocular that can see the end the universe, so he put it in a tripod and bend over to adjust it. Looking through the viewfinder he was supposed to see the very end of the universe. The distances in the universe are enormous - the more he looked, the more he was convinced that he was looking at some pants. Slowly it ~~dawn~~dawns to him that he was looking at his own behind. And this is one more proof that the universe is spherical.

WHAT WE SEE IN THE SKY "NOW?"

The light travels with 300,000 km/sec, this is real fast, but the distances in the universe are so enormous that this speed actually is real slow. We receive our "information" about certain events from hundreds of thousands to billions of years after these events have happened. And once we receive the "actual footage" of what has happened, let us say, 5,000,000 years ago, the stream-of-events go minute by minute; back at that time. So we are watching actual footage of events that one of them has happened 5,000,000 years ago, but next to it we are watching minute by minute what happen to a star 5,000 years ago. So anything we watch is minute by minute. The question is how much time it took until the footage come to the "screen" on our sky~~-~~? Let us look at the footage from all the stars in our Galaxy that has happened between 100,000 and 4,000 years ago. In the same time we see the near-by planets and moons from the Solar System, from which the most immediate is the footage from out Moon coming in 1.3 sec; from the closest star Proxima Centauri 4.2 years delay; from Andromeda Galaxy 2.5 million years delay. So we are watching numerous "movies" on the sky, but all these minute by minute shows are from totally different times and locations. Once all these "news events" pass through our sky we cannot "rewind" the "tape." What have happened before the actual footage came to us, or what would happen in the distant future is unknown and subject of speculations and assumptions. The case is different if some "events" that we have watched for years and have record of them. We might be good at speculations and assumptions, but the question is how good we are at discovering the truth about the universe? And you know already my opinion that ~~90~~85% of the accepted theories are 100% NONSENSE.

Here is another aspect of what we are seeing right now, and what we will never be able to see about our Galaxy and the stars in it. Let's say that arbitrarily our Galaxy has existed in the past 13.8 billion years. Reportedly, our Galaxy is 100,000 light years across. What I am trying to say is that all the "information" about our Galaxy and the stars in it is less than 100,000 years old. Our Galaxy's age is 13.8 billion years, and in this case information of what happened in the prior to these 100, 000 years could not be available to us ever.

Let's see in what "world" the leading scientists are living right now. For them all began with some nonsense called Big Bang, so they look through the prism of this event and see the following deceptions: Some "star-creation" and "evolution" since the mythical Big Bang. Classification of stars through "luminosity" and fake "surface temperatures" – all this is 100% garbage. Stars are filled with gas - this is one of the big lies coming from one of the friends of the biggest liar ever Einstein - Arthur Eddington. Read this quote about him from Smithsonian Universe 2020 page 251 "He studied the internal structure of the stars." Here is my question, how he could have "studied" this "internal structure" when no such information is available? Maybe the Little Riding Red Hood invited him inside one of the stars where Eddington saw with his own eyes the nuclear reaction going on at its core? He "also calculated the abundance of hydrogen in stars" – what an idiot! He also calculated how many proton and neutrons are in the universe without knowing its size! For his nonsense he was knighted in 1930, and now you can understand how the stars got filled with hydrogen gas instead of chocolate. What he forgot to calculate is how many beetles can enter in his ~~dumb~~ head. In another textbook I've read that the scientists have been "studying" some star in the last 400 years. How they can "study" something when they do not know its internal structure?

I am of the opinion that from the point of view of the Grand Universal Cycle we are in the middle of the local "crunch." These two Magellanic Clouds and other small galaxies that at the edge of our Galaxy are here in preparation for the "local" "big" crunch which would come to a full blow when Milky Way collides with Andromeda which is surrounded with other smaller galaxies getting ready for this event. And we can say that this "gathering" of the local galaxies is prescribed by the "grand universal design."

Are you ready for another joke? Here is a short story about why one student became an astronomer. The future astronomer was talking to his fellow student. "You know I am not so good at mathematics, and I wonder in what scientific field I should major ~~in~~?"

"If you are not good at real math, there is a fake one in the science of Astronomy, and there you can write all kinds of fake formulas, and even formulas that only you can understand. As a matter of ~~fact~~fact, the more idiotic your formulas are - the more famous you can get, and more prestigious jobs you can take. ~~Finally~~Finally, you'll be the only one understanding the universe and the rest of the people have to listen to your fantastic nonsense in order to learn about it."

"This sound fantastic – that is exactly what I wanted to become!" And then he become very famous astronomer and wrote many books that people all over the world read his nonsense, and that is the reason why today people do not know much about galaxies, stars, or the Earth. His name could have been Stephen Hawking. But here comes the real test - the Earth is in trouble right now, and soon people will realize that the established scientists are only milking the society for $ without delivering any real knowledge or understanding.

THE HISTORY OF SCIENTIFIC DEAD-ENDS

Quotation from The Universe by Hoffmann 1994a textbook: "We know that the stars and galaxies are made of substances found here on Earth and that all atoms, regardless of their locations in the universe, obey the same physical laws. As a result, laboratory experiments conducted here on Earth can provide the insight and understanding needed to probe the distant reaches of the universe." Firstly the stars and galaxies are absolutely not made of substances found here on Earth.! For this reason whatever the scientists do in their labs, they probably will never be able to reproduce black sphere substance of which all black spheres; star; and the cores of brown dwarfs; planets and moons are made of. Basically this quote is absolute crap! Here on Earth we are surrounded only by "matter # 2" which consists of atoms and molecules, and this "matter" is about 0.5% to 3% of the total mass of the universe, so all mathematical and physical "laws" we have discovered and know – do not apply to the rest of the universe, except for gas, dust, and partially for planets and moons.

Some planets and moons are alive which means that they have working black spheres inside. Those that we would call "dead" no longer have working black sphere inside. The Earth has "working" black sphere inside (the so-called "solid iron core") which is made of the same substance that the Sun is made – black sphere substance. The rest of our planet is made of "matter # 2" which is atoms and molecules – this is the metal, the mantle, and the crust we live on, as well as the atmosphere. On the other handhand, Mars and our Moon are "dead." Initially at formation, they had live black spheres inside, as there is no any other way of creating planets and moons, but now their BSs have "died" and "disappeared." So Mars and our Moon are made only of matter number two – atoms and molecules. The dead black spheres are turning to iron, or to iron and stone.

The story goes that the Greeks realized that through mathematics and geometry the universe could be discovered, and this is true to certain extend. So they did discover quite a lot. Then there are more discoveries in the field of chemistry, physics, and mathematics in the last 500 years, but the wrong assumption that the universe is made of atoms and molecules leads to substantial misconceptions. Yes, there was advancement in the human knowledge about the universe, but there wereas a lot of deceptions as well, due to not knowing that there are three kinds of matters in the universe. Stars are the most abundant in the universe, and humans decided that they should be filled with "matter # 2" namely with hydrogen gas. So the deception of the human race continues with hydrogen-filled stars. I do not understand how the entire scientific community believes this nonsense without anyone to see this gross misconception.

The fact that the scientific community can get itself in so many profound misconceptions is disturbing. What went wrong with the "human race's reasoning?" How we become so gullible? I see a mistake in the "authoritarian" imposition of shire "speculations" as acceptable "truth." Terminology like "leading theory" for the Big Bang is like "imposing" "stupidity" on all people whether they are scientists or not. If we ask who is the leading bicyclist in the tour of France? We could know exactly who he is, but how we could determine which is the "leading nonsensical ~~speculation~~theory?" The final result unfortunately is "mass deception." The pattern of the deception has been going on for more than 300 years beginning with Isaac Newtons' nonsense, and that was the beginning – there were many more to follow. ~~But~~ If we acknowledge the fact that the human race is not that smart; we could become better ~~ACKNOWLEDGE this unfortunate FACT, we COULD become BETTER~~ human beings. Human race repeats the same mistakes to perpetuity, as if it is unable to learn from them. ~~Yes~~Yes, we are smarter than monkeys, but what that supposed to mean? I will go farther - we are way-dumber than them. Monkeys do not seek world-domination, monkeys do not enslave other monkeys, the list can go on forever – we are the ultimate monsters. Bigger monsters than us do not exist and have never existed before. We live in the grace of some precarious circumstances - in the middle of some nuclear reactions. These nuclear reactions do not necessarily have any special safe place for us to live; ~~therefore~~therefore, we are at the mercy of the circumstances. Instead of having love and tolerance toward each other, we have continuous strife for domination, dehumanization, exploitation, subjugation, manipulation, and extermination. At this very moment we are in precarious situation where complete annihilation of the human race is possible coming from outside.

The "intruder" is "removing" asteroids and comets from their "safe for us orbits," and also is capable of "killing" the planet Earth, because the evidence are clear that Mars' living atmosphere was "killed" this way, and as scientific investigator only I know how that happened. Today the attack on our Earth is similar to what had happened to Mars 700,000,000 years ago, and kill its living atmosphere. Let us look at the fact that one of the Uranus' poles points toward the Sun. Scientist's interpretation about this is that it was caused by some collision, but I know that is caused by outside influence of large size BS which at certain time came and challenged the order established by the Sun. So these "disturbances" to the planets in the Solar System come and go. The current disturbance of all planets; moons; and rocks is going on right now. Would our fragile mother Earth get through this "onslaught" "unscathed?" And if it does, we have better chance for survival? Courage children, the fairy tales from mom and pap, might not be applicable for the "new" requirements for "survival." The word "survival" might get a new meaning; like continuation the existence of the human race. The severity of what we have to go through, looks like, depends on some external celestial body, and how close it would get to the Earth. We cannot do anything against the intruder, because it is too powerful, but we can get united and create protection against asteroids and save foods. Many times the evil side of the human race manages to win, because of our dishonesty, greed, lack of transparency, freedom of speech, common stupidity, etc. … Are we doomed now? In the final analysis, if we do not acknowledge more-than-obvious fact that we humans collectively are way-more stupid than what we are ready to admit; we are going to continue governing ourselves improperly; and our "science" would continue to be ruled by the political mafia. It is time for a change. Change or perish. The survival of the fittest; how about the survival of honest scientists and all?!

Here are two erroneous assumptions that still govern our "thinking" and "reasoning:" The first one is REPUTATION. Einstein, who is a common moron, acquired "reputation" after shooting his mouth that light is bended by the gravity. By the way, this complete nonsense is still accepted as a "truth" today! After acquiring the necessary reputation, the little MATHEMATICAL CROOK milked the world to the rest of his life – feeding the hungry for knowledge human race about the universe with nonsense like "singularity;" "gravity bends the fabric of the space around it;" e = mc2; time throughout the universe is different; and two theory of relativities that are COMPLETE AND TOTAL NONSENSE! But hey, let see what the contemporary scientists think about this? They accept his nonsense … today!!! Are we getting smarter, more stupid, or what? We are getting smarter – give it 100,000 years or more! My duty is to speak-up, and you can believe any one you want. Looks like that the "good reputation" of some scientists is like a "Trojan horse" for passing deceptive ideas. Many of today's deceptions in the field of Astronomy are done by scientists who have some real contributions to science like Lorenz and Plank. Isaac Newton wrote a book on gravity which extends to other celestial bodies. The problem with his "discovery" is that every dog knows that if something is thrown up; eventually would go down. Beyond Roche's limit of falling objects this "gravity" does not extend to other close-by celestial bodies. Newton has made two enormous errors: the first one is that he did not understood Kepler's third law and "improve" it through DESTROING it, and the second is his "convenient assumptions" that planets should move with some "even" speeds so that his "no existing gravity" could "capture" them. Even for falling objects to the ground his gravity applies only to falling apples, but I doubt that applies to falling iron (no experiments have been done~~made~~ yet). We should not blame only Newton. Is the rest of the scientific community like some sort of small children that listen and believe him as if he is their father? Let's say that the first scientists back then made the mistake accepting his theories, but here we are in twenty first century with the sky crowded with all kind of satellites, with good telescopes, with universities (spreading lies) equipped with labs, yet the yesterday's nonsense still stand! How many scientists and mathematicians since Newton's garbage came 300 years ago used his mathematics and no one saw this outrageous mistake of misinterpreting Kepler's third law?

A leading book says that Eddington only "suggested" that stars are filled with hydrogen gas. Here I am an investigator; imagine also that Eddington is alive right now and I question him~~:~~ ": "Mr. Eddington, why did you lie to the world that stars are filled with gas?" Let's assume that his answer is~~:~~ ": "I did not lie to them, I simply SUGGESTED." If he only "suggested," he is innocent. Whom should we blame then? This raises the question, who is in charge in the scientific community accepting so many lies and legitimizing them? As an inspector and new in the field, here are my findings: There is enormous amount of "scientific trash" piled-up through the centuries – time has come for a clean-up! I am the only one in the world, who knows the real reasons for global warming and weather havoc, but I am out of the loop, I need to belong to the scientific community, HELP!. After I discovered the universe, and realized that the global warming is not solely human-made, I wanted to talk publicly about it. First they asked: "Who are you?" Turn out that I am nobody. "What did you work through your life?" The answer is small-time builder. A scientistsA scientist is not the one who gets the fake Astronomical education (90% deceptions); but the one who has inquiring mind; SEEK THE TRUTH; AND COULD REASON. If Galileo Galilei was put under a house arrest and prohibited from broadcasting his findings, today the human mafia requires peer-review. Google, You Tube, and Wikipedia are spreaders of lies and misconceptions. Christ was killed, and after that millions of "Chists" were murdered; whether physically or spiritually. And the human ruling mafia also is "killing" the quest for knowledge!

Here is a quotation from a textbook: "Astronomy is built on the understanding of light, and every scientific discovery about light has lead to important discoveries about the universe." "I scratched my head, and every scratch lead to new scientific discoveries, therefore I have to scratch my head more often." There is no difference in the meaning in these two statements. The phony reasoning goes farther: Allegedly the light emitted from the stars determines their temperature. THIS IS A LIE! The truth is that this temperature is created outside and has nothing to do with the temperature of the star itself. All their reasoning sound quite reasonable on the surface to fool the "true believers." The problem is that stars are not made by any atoms and molecules, and what they are looking at are gases and matter formed outside of the stars as the stars release BBPA and the nuclear reaction of self-formation of atoms takes place. There is a correlation between elements that are present around certain stars and the temperature generated by the nuclear creation, but has nothing to do with the substance of the stars and its temperature. This process of self- formation of certain chemical elements outside of the stars cannot in any way show what the temperature of a star is or what the star is made of. Can they be more fartherfarther from the truth?

The stew of misconceptions is endless. It is a chain of deceptions piled through the centuries. Falsehoods like nothing can travel faster than the speed of light, when xX rays, gamma rays, and neutrinos do travel faster. Here is some "deep thought:" "The stars are almost perfect black bodies." In reality stars are not "black bodies" at all, but from this assumption comehere are some laws like Stefan-Boltzmann's and Wien's laws. These laws apply for the co-called "black bodies" temperatures which in lab conditions supposedly "resemble stars." BALONEY!!!

The thought behind this~~at~~ is that "we can discover the stars through creating similar conditions in labs." Sorry to break the celebration, but no matter what you recreate in your labs; you will never be able to recreate what is in the stars – black sphere substance. So, readers, you have been informed how the today's scientists are "investigating" the stars.

Here is how erroneously they measure the temperature of the Sun. Through satellite orbiting the Earth, they have measured the energy the Earth receives from the Sun and it comes to 1,370 watts per square meter. Supposedly this temperature comes from the Photosphere which has the temperature of 5,700 C, but the temperature of 1 to 2 million degrees C of the Corona which extends several million kilometers from the Sun does not come to this equation. Why this is happening? Because they cannot explain the corona's temperature! The temperature of 5,700 C in the Photosphere from the distance of 150,000,000 km CAN WARM the Earth to around minus 260 C., but this would ruin their "luminosity" NONSENSE.

Allegedly a nuclear reaction is going on inside the Sun which is filled with hydrogen gas. If this is the case, the Sun should not have any real "surface" according to their reasoning, but the surface of the Sun is clearly visible! THEY REFUSE TO SEE IT!? So currently according to them, the Photosphere IS THE SURFACE OF THE SUN! BALONEY! Keep in mind that these people for years are "studying" the Sun with satellites! The real surface of the Sun is clearly visible through the so-called "black spots." There is X ray picture of the Sun from August. 21. 1973 – on this picture absolutely and unequivocally the curvature of the black spherical surface of the Sun is VISIBLE! Why they do not see it - because it might ruin their "black body" nonsensical theories and some of their careers with "we need the money" mentality.

"Stars are classified according to the appearance of their spectra in a way that reveals their temperature." Then there is Herzsprung – Russell diagram where a regular pattern apparently appears when the absolute magnitude of the stars is plotted against their color indices. From there comes the terminology about "main sequence stars" to which the Sun supposedly belongs. Let's say that you have to investigate the "scientific" details, but these details are one enormous sewer pit! That's where we are right now. All these "scientific" claims are backed by substantial volume of formulas, discoveries, and scientific "integrity" like that of Isaac Newton, Einstein, and many others. Now the scientists "know" what processes are going inside the stars, and with what chemical elements they are filled. This is the current condition of "reasoning" in the "dark ages" of today! Why some stars like the Sun appear yellow-red and others white, is because the slower speed of release of BBPA of the Son allows the creation of this Photosphere which is the initial formation of more compound atomic particles, where bigger stars have more energetic release of BBPA and the eventual formation of a Photosphere becomes impossible. Another words – it is blown away.

EXAMPLES FOR FAKE PHYSICS AND MATHEMATICS

I am trying to ascertain when and how the science of Astronomy ~~get~~gets into this unbelievably high number of nonsense (90%), and at this moment I am of the opinion that things get in "wrong track" with introduction of mathematics in it. Here is Newton, who introduced mathematics. All that he says is 100% wrong. After Newton more scientists rely on physics and mathematics, and the pile of mathematical nonsense become larger and larger. Now, 90% of all math, formulas, tables, and classifications have to be discarded. After Newton scientists began writing all kinds of "laws," as if they should order the Universe how to behave, and the observations should take second seat.

Here is how the Sun has to obey some "mathematical requirements." Let us look at Bode's law about some numerical sequence about the orbital distances of the planets in the Solar System. This was a real law, because shows the real orbital distances, but it was not accepted by the scientific community because it did not "satisfy" some abstract math requirement of "N." Hello, wake up, the Sun is not interested in your requirements! As a ~~result~~result, the human race is not informed properly because their scientists have some dumb math requirements! So the Astronomy is no longer based on observations, but on dumb theorems, physical laws based only on matter number two (3̶2% of the total), and when you add the professional liars and careerists like Einstein, Eddington, and cronies like Hawking who does not know anything but wrote book after book, and you got the picture of the current mess.

I have made the statement ~~in this book~~ that the science of Astronomy consists of 90% "scientific" garbage. If I take one of their textbooks and put it through analysis, I have to write a boring book of 1,000 pages. Here I want to give you just a small example of their scientific nonsense, and keep in mind that I just opened one of their textbooks at random, and it's like looking for trash in a trash-container. Bear with me, it has to be done, otherwise they will continue to pretend that their claims are backed by some kind of "science," and I am just a fool who does not know what is talking about. "During hydrogen burning, a portion of the star's mass is converted into energy." Two fundamental mistakes in this sentence: (by the way, "burning" means nuclear reaction where hydrogen becomes helium initially in the core of each star). Firstly, stars are black spheres packed with basic building particles of the atom, and no atoms of any kind can be inside them, this "burning" is figment of Eddington's "guessing." He has never been inside a star, so how he could "know?" Matter is never converted into energy, and this is another example of their erroneous assumptions. The universe has existed forever, and that leads to the only logical conclusion: Not the smallest amount of energy or matter is ever lost! The quotation continues: "We can use Einstein's famous equation to calculate how long a star can last." We are talking about $e = mc^2$. Notice here this "reverence" with the word "famous". Maybe this nonsensical formula should be chiseled in stone, or put in some supplement to the Bible, and if the universe does not obey it – it is universe's fault, but the best thing to do with it is to put it in its rightful place – the trash. Matter and energy go hand in hand, they coexist, and when energy is released, it does not mean that matter is lost. Energy is not "wasted;" it is just "present," and would be "present" forever.

Let us see some other fake formula E = fMc2 (2 means square), where "f" is the fraction of the star's mass that would be converted to energy. Then another fake formula is introduced E = Lt. Here "L" is the ~~stars~~star's luminosity and "t" is the time hydrogen will burn into helium. Scientists like the word "luminosity" because they think that this is some sort of indicator. They see luminosity, characterize it, and make their "scientific conclusions;" which are "scientific baloney." From these two equations they arrive to this formula Lt = fMc2. It is only logical that from two nonsensical formulas we could only arrive at a new compound "nonsense." And that is how they calculate how long a star will stay in the so-called "main sequence." So when they tell you that our Sun has 5 ~~billions~~billion more years of life – UNDERSAND THAT THEY DO NOT KNOW 100% WHAT THEY ARE TALKING ABOUT! My message to Einstein's and Hawking believers is: You can and you have fooled the entire world, but you are not going to fool me! The list of nonsensical formulas like this one can take ~~hundred~~hundreds of pages.

DETAILED EXPLANATION OF KEPLER'S THIRD LAW

A disclosure, I have never been a mathematician, but here I am finding mathematical mistake in one of the biggest – Isaac Newton. Kepler's third law states that if one knows the distance a planet's orbit to the San, can calculate the time of its orbit. Let us not forget that orbits are elliptical, so the distance is averaged. Now I will prove to you that both numbers of the orbital distance and the time for one orbit for the planet Earth ~~make,~~ must be "one" (1.0). ANY SUBTITUTION WITH ANOTHER NUMBER – RENDERS THIS FORMULA INVALID! The Kepler third "law" states that the orbital distance cubed equals the time of the orbit squared. If (a) is the average distance to the Sun, and (p) is the time of one orbit, then p2 = a3, but ~~this equations~~this equation for all planets are accurate only if both of these measurements for the Earth are taken as (one) 1.0 and NOTHING ELSE! The "great" mathematician Isaac Newton FAILED to understand this key provision – ENORMOUS MISTAKE – he substituted the distance of the Earth from "1 AU" to meters, and "1 year" of orbit to seconds!!! Here are his numbers: does this equation seem accurate to you? 31,560,000 seconds SQUARED should equal 149,600,000,000 meters CUBED. It is obvious that this is impossible. Pretty dumb, isn't it? Dumb; dumber – Newton! But what should we say about all these scientists and mathematicians who in the last 300 years FAILED to discover this mistake? Today this nonsense is probably in some supper computers. With this colossally inaccurate formula the Sun's mass has been CALCULATED. Now let us look at comparison density between the Sun and the Earth. ~~a~~According to the current scientific "findings" (density means weight per cubic meter): For the Sun is 1.4 tons, and for the Earth is 5.5 tons. Let me give you the real numbers for the Sun calculated with some approximation – Sun's density might be about 2,365 tons per cubic meter!!!~~.~~

MASS AND DENSITY

Basically we have three fundamentally different masses which have substantially different densities. The first is "matter # 1" in all BSs which changes its density through time, because of constant release of BBPA. In comparison "matter # 2" is very light where every atom consists of core where almost all of the atom's mass is located, and the electrons orbiting this core at some enormous distance compared to the size of this core. That is the reason why "matter # 2" which consists of atoms is substantially "lighter" compared to the "matter # 1." Third is "matter # 3," which is compacted "matter # 1" found in "white dwarfs" and neutron stars. The Sun which is 100% BS has density of 2,365 tons per cubic meter right now after 13.8 billion years of releasing BBPA. Allegedly, if our Sun is to become a "white dwarf", which happens only in supernova explosions, its mass has to be squashed into the size of the Earth.

Things get trickier, because scientists give two different numbers of comparison the size of the core to the size of the empty space in a~~an~~ atom. ~~for the size of the core of the atom compared to the empty space in each atom which is determined by the distance of the orbiting electrons.~~ These numbers are 10,000 and 100,000. According to these numbers the Earth could be shrunk to 1.2756 meter with first number or 0.12756 meter using the second number. I doubt this numbers; they look exaggerated when comparing the density of the Earth's 5.5 tons per square cube and the Moon's 3.3, when the Earth has live BS inside and the Moon do not.

Let us summarize these three densities being the most important in the universe. The lightest is "matter # 2," which consists of mostly rocky and metal material of our Moon with density of 3.3 tons per cubic meter. Because the Earth has both matters #2 which is the mantle, crust, and atmosphere, and matter #1 in the BS; its density is higher 5.5 tons per square meter. The Sun is 100% BS which is pure matter #1 has density of 2,365 tons per sq. m. "White dwarfs" and neutron stars probably have the highest possible density, and it is 1,000 times greater than the Sun's or any BS.

At this moment the size of the Earth increases~~, its size,~~ but its overall density almost ~~is getting smaller~~does not change. This ~~is~~ happen~~sing~~ because the BBPA create atoms and the planet gets bigger in size, but in the same time the BS's density is getting smaller. The amount of BBPA coming out of the BS equals the amount of BBPA that creates atoms and molecules (matter # 2) but some of it becomes gas and goes into atmosphere. I suspect that the Sun also increases the size of the Earth but very insignificantly, and experiments have to be conducted.

As you already know I have new definitions for planets and brown dwarfs. These new definitions are way more scientifically accurate and very important for future space travelers. When we characterize the moons we have to know that they can be only planets or brown dwarfs. Most of the moons in the SS are planets, but Io and Titan are definitely brown dwarfs. Brown dwarfs (BD) are the celestial bodies between planets and stars. Being more powerful than the planets; they begin to reject "matter # 2." Their BSs are bigger than 1,220 km diameter, and with their bigger BSs their "rejection" of matter #2 gets stronger and stronger, so their density increases. The biggest size BDs shine "almost" like stars. Venus is definitely the smallest possible BD in the universe, because celestial body with smaller BS inside happens to be the Earth, which is the biggest possible planet in the face of the universe. Venus is very interesting in this categorization because it is right at the border line between planets and brown dwarfs. Here is the new definition for a planet: Planet is a celestial body on which a spacecraft could be able to land, and astronauts could be able to walk on its surface (planets and moons are the same). Keep in mind that living atmosphere is possible only on planets and never on brown dwarfs. On Venus with the surface temperature of 480 degrees C, walking on it is not advisable. No one has the information what the size of its "solid iron core" is, but I know that is bigger than the Earth's. From this "info" we could draw the conclusion that Venus density is bigger than the Earth's. The scientists "guessed" it and they guessed it wrong! Density increases with the next "brown dwarf" after Venus which is Jupiter's moon Io. Why Io has smaller outer size than Venus, but is more powerful, because it has bigger BS that rejects more "mater # 2." Io could had poisonous atmosphere like Venus, but because is too close to mighty Jupiter; loses it instantaneously. Io is more powerful than the Earth, but it is a moon; that means that our Earth could have been Jupiter's moon.

The four gas giants are all brown dwarfs, and from them the most powerful one is Neptune, and the weakest is Jupiter. Jupiter has the biggest size on the outside which is just too much gas. From its radius 60,000 km is gas, then at his diameter of 143,000 km let5 us deduct 120,000 km is gas; what is left as solid ball is 23,000 km (twice the Earth's size), but it has the biggest magnetic field (20,000 times bigger than the Earth's), because his "generator" is the biggest in SS. I challenge the scientists' belief that magnetic field can be created without iron. Listen to the erroneous explanation from "Smithsonian Universe" 2020 page 200 about Uranus which is more powerful than Jupiter (my classification): "It is made mainly of water, methane, and ammonia ices, which are surrounded by a gaseous layer. Electric currents within its icy layer are believed to generate the planet's

36

magnetic field." "Smithsonians," I am already beyond the terminology of "believing," I am simply telling you what it is. Jupiter's magnetic field is 20,000 bigger than Earth's, when Saturn barely has a magnetic field of 0.7 from the Earth's, and that teaches us an important lesson: Brown dwarfs with black spheres inside them which are bigger than Jupiter's lose their magnetic fields! Why this is happening? Because with the faster release of BBPA Saturn's BS no longer "tolerates" iron on its surface, and as a result of this cannot have a magnetic field. Reportedly Saturn has a moon orbiting at 25,000,000 km – talking about power, and notice that no magnetic field is needed for that.

By the way, "Smithsonian Universe" 2020 reports that Jupiter's "iron core" is 10 times the size of the Earth. This is probably the entire solid body of this planet, but we are interested in the size of the BS only. Another source is giving me the size of Saturn's "solid" core to be one and one-halve larger than Jupiter's. So, no reliable data is given, and no reliable conclusions could be made, but I would continue to claim that Saturn is stronger than Jupiter.

More glory to our beloved BS, as we are alive because it is alive. Thank you, mother black sphere for sheltering us with your magnetic field for our atmosphere, and thank you for the hot lava that keeps the interior of the planet hot so we can have "ground" and "underground" water without it sinking into the interior. Thank you for having moderate volcanoes; so we can breathe. We also have to thank another BS, and that is our Sun, doing outstanding job of keeping the Earth warm for 4.6 billion years, so that we can get evolved from frogs and rats to the today's advanced stages of "smarter" than monkeys. And let us not forget the contributions of the space aliens for our development. I am sure that in the next 100,000 years the human race might reach the high morality of our DOGS.

Sun's mass is denoted in kilograms – so is the weight of cheese on Earth, and on another planet one kg weights differently. How the measure of "kilograms" can have any "universal" value and apply to other celestial bodies, except for commercial weighing here on Earth? So, firstly, Kepler's third law is used improperly to measure the Sun's mass, second the mass of the Sun is denoted the same way like cheese on the Earth. There is a third "political" and unscientific reason the density of the Sun is given as low as 1.4 tons per cubic meter, when compared to 5.5 tons/cubic meter for the Earth. The story goes that Eddington only "suggested" that the stars are filled with gas. The scientists are like fish that took the bait, and as a result this nonsense is taken as a "reality." How low the human race can go? Among other nonsense, notice the scientists' vocabulary of self-aggrandizement through labeling some erroneous propositions as "laws." So, Hubble's "constant" of expansion of the universe for example becomes Hubble's "law," but the fact is that his "law" is complete nonsense - who cares?! All of Newton's "laws" are absolute baloney, let us look at his third "law": "Whenever one body exerts a force on a second body, the second body exerts an equal and opposite force on the first body." This is complete nonsense. The bigger BS "commands" the smaller BSs, and that is all, there is no such "equal and opposite force." The universe is "command and control" structure based on size of the BSs. These ~~are~~ SYSTEMS are based on th~~ee~~ size of the BSs; the bigger commands the smaller. I challenge the notion the Moon exerts any influence on the Earth. They claim that affects the water in the oceans, which I doubt – I have to conduct my own investigation. The Earth "commands" the Moon to be at this orbital distance at this predetermined speed. The "command and control" structure does not allow any increase of the orbital distance, and the reported 3.3 centimeters increases per year are nonsense! On a separate note, the Moon has to orbit around the precise center of the Earth, and not at some imaginary point inside the Earth because supposedly it is binary system – it is not! The Earth has full command and control over~~f~~ the Moon. The funny thing is that at this very moment our BS is off of its center – this particular investigation should be postponed for 200 years or so, when the "intruder" leaves the Solar System, and hopefully our BS returns back to the center of the Earth. Scientists wonder how our planet can handle such a big moon. What counts is the size of the BS inside. The Moon has lost its BS, so it is 100% matter number two – atoms and molecules, but the Earth has its BS working therefore consists of both matter number one and matter number two.

NEUTRINOS

Scientists wake-up, neutrinos travel faster than the speed of light! When this fact would be acknowledged?! When a neutrino hits a proton in the water-filled drum a positron is produced, and brief flash of light follows, and this is known as "Cerenkov radiation." He was a Russian physicist and observed this 1934. The neutrinos were moving through water faster than the speed of light, but somehow the "scientific politburo" decided that in "vacuum" nothing can be faster than the speed of light. I wonder what the opinion of the mathematical crook Einstein was back then, because his moronic theories are based on the assumption that the speed of light is the speed limit of the universe. I am not quite sure, but today's scientific mafia uses widely the moronic assumption that the speed of light is some sort of "speed limit." Their premise that light remains the fastest in vacuum was crushed again in 1987 when neutrinos arrived to the Earth 3 hours before the light arrived. On top of it their inconsistent reasoning is that if neutrinos do not have any mass they would travel with the speed of light, but they reportedly have some real small mass, and still travel faster than light! In this "competition" in 1987 of reported distance of 168,000 light years, neutrinos beat the light with 3 hours!!! In 2011 in Switzerland again was recorded that neutrinos are faster going to Italy. No one wants to overturn the boat and acknowledges the truth because "e = mc2" and many other fake formulas containing the speed of light HAVE TO BE TRASHED. Tell me if this is not a mafia!?

HOW WE CAN LEARN MORE ABOUT THE BLACK SPHRES?

We would never be able to go inside black spheres, but we can observe what they do and learn from it. For ~~example~~example, they change their appearance and internal structure depending on the different tasks they perform, and let's never forget that they "live" and "exist" forever. The only ways to learn about them is through collecting unbiased data, and then separate the actual data from conclusions and theories. Eventually, hypothesis and theories could lead to knowledge, but at this time we have erroneous theories and conclusions that preclude any real discoveries, and the advancement of new hypothesis and theories is impossible. Reading through "Smithsonian Universe" 2020 I wanted to know: Are there any live volcanoes on Mercury right now? For all the volcanoes they write that they are under 500 years old. For them this is some sort of "scientific information," but for me is not. There are many such examples where the nonsense coming from their theories takes precedence over the actual facts. At this moment the scientists are accepting some leading theories as if they are the very "reality" and are desperately trying to proof them. When this "house" of "phony theories" and "laws" would collapse!? Looks like the thought "We might be wrong?" cannot pass through their heads. Not only that they have been wrong, but at this moment through not understanding the climate changes - it borders on a criminal behavior on their part. As a matter of ~~fact~~fact, it is. They are more interested on the perks, the salaries, and TV shows than in finding the "scientific truth."

At least this book tells you what the BSs are, and what they do, and of course the quest for new discoveries will never end, but only through the right theories. If UFOs are breaking down and falling to the ground, what that tells you? Despite that the space aliens are way more advanced civilizations – their UFOs cannot pass the safety test! Then, when the quest for more and more knowledge could end - never!

We and the space aliens are biological last stage intelligent creatures, but they are way ahead of us. They play the same games with us, as we ply on the other animals, as we put colors with transmitters on them, and they are giving us knowledge when we sleep, and so on and so forth. They can cure us from disease, bring to life prehistoric animals, make new animals from different parts, and ~~may be~~ ,may be, bringing long-dead people to life as well.

OBSERVATIONAL SCIENTISTS ARE REAL, BUT COMPUTERS MATHEMATICIANS ARE FAKE

I have to mention something about myself. I consider myself "observational" scientist like Copernicus, Tyco Bra~~h~~e, Kepler, Galileo, and Bode. These were the real scientists. Here is a quote from "Universe" 1994 by Kaufmann, page 66:

"Until the mid-seventeenth century, virtually all mathematical astronomy was entirely empirical, characterized by trial and error. From Ptolemy to Kepler, essentially the same approach was used. Astronomers would work directly from data and observations, adjusting ideas and calculations until the right answers finally emerged." Glory, glory, alleluia! Here is what the first "scientific" servant of Satan did (from the same page 66): "Isaac Newton introduced a new approach. He made three assumptions, now called Newton's laws of motions." Newton's assumptions are all wrong, detrimental, deceptive, and Satan-serving! The human race bought it, and now is paying the price with its ignorance about the universe. I cannot hold myself, but mention here the "ultimate sewer of the human scientific thinking" – Einstein!

SOME THINGS THAT THE BS DO

Here are some things that the Black Spheres perform: Firstly, they release BBPA at rate proportional to their size. Second, they dismantle any atom that falls into them to BBPA. Third, they have this "field of influence" which extends outwards depending on size/strength. Different sizes BS perform different tasks. There are no uniform characteristics for the different sizes. Most definitely the bigger are more powerful, but we cannot expect uniformity. Most of them have this influence around the equator where we see "rings" around the 4 gas giants, and in this field are the moons rotating around the planets. One planet has to rotate around its axis otherwise cannot have moons. Mercury and Venus are such examples, so we can have the rule: "If a planet is not rotating therefore cannot have moons." Obviously the Sun prohibits the rotation of these first two planets, and all moons do not rotate, and subsequently nothing may orbit around them. The name "System" for the Solar System is quite appropriate, because it is a system within a bigger system, which on its term is part of even bigger system, but inside it are the smaller systems of the planets and their moons. I suspect that the "role" of the Sun could be "performed" by brown dwarfs, but that system would be definitely smaller. For such a role the BD should be of substantial size. Neptune has a moon orbiting at 48,000,000 km.

We have to address the strength and the influence of celestial bodies that are not part of the Solar System, but somehow they have entered "unwelcome" in it. And to me the most "dangerous" are the "white dwarfs" where the power of the Sun can be stashed in a planet with the size of the Earth, and subsequently difficult to see and detect. If we want we could know what influences the Earth receives, which would be similar to the influences of the Sun. There are three influences I can think up right now, and they are: first trying to "push" the magnetic axis perpendicular to itself (the North Pole moving from Canada to Siberia); second is the "attempt" to make the mantle rotate around some different axis from the "established" one with a different speed; and the third is to redirect the Earth to orbit the intruder. Here we observe something surprising and rather disturbing; there is "different" treatment of matter # 1 (BS) and matter # 2 (consisting of melted iron, mantle, crust, and atmosphere). The result is displaced BS. As a result of this some wobbling is happening as well.

Let us examine some "influences" a strong celestial body could have on a weaker one (the weaker being the Earth). Instead of pulling the weaker body straight to itself; as the Newton's gravity claims; the stronger one "directs" the weaker to go around it. Mr. Newton did not understand that, and that is the reason why his mathematics were wrong, and if you do not believe me, ask the "artificial intelligence," where the stupid become smart. This influence is not understood by today's scientists. They call it "gravitation," and because of this, they use the inaccurate mathematical formulas of Isaac Newton; instead to find the real formulas. This influence is like multiple spherical layers resembling onion peels. This influence is all around certain celestial body. This particular influence of the Sun handles Pluto, and for that matter would handle any smaller body regardless is it inside the SS or is coming from outside. Looks like the Sun preserve the speed of this body and only redirects it to go around the Sun. -Around the equator of the Sun there is a flat field where all planets are orbiting. The number of planets "preordained" to orbit the Sun is 9, but the 5th orbiting place is empty. Much the debris in the SS were captured there That means that if small celestial body enters the SS in this field with a zero angle, then it would be subjected first to the inevitable "order" to go around the Sun, but when crossing any of these orbital distances, it should be subjected to even stronger "order" TO STAY IN ANY OF THESE ROTATIONAL ORBITS. That is how all these asteroids end up "captured" in the empty orbital space between Mars and Jupiter. Other aspect of the Sun's influence is this: THE CLOSER THE PLANET IS TO THE SUN; THE FASTER IT HAS TO ORBIT. Mercury's orbital speed is 47 km per second, when the Pluto's speed is 4.7. Although Neptune is the slowest; it is the most powerful. The last influence of the Sun, as an example of influence on smaller celestial bodies is the PROHIBITION OF ROTATION AROUND THEIR ASIX OF THE TWO CLOSEST PLANETS; namely Mercury and Venus. If a strong white dwarf enters the SS we know how the Sun would handle it; the question is how its strong influence would affect the planets in the SS? All celestial bodies have influence which comes from the amount of BBPA in it. This influence is ever-present; it cannot be "canceled," modified, or disappear. If we know what is the influence of only one Basic Building Particle of the Atom, and know how many of these Particles are in certain celestial body, we could know exactly its influence. All brown dwarfs in SS handle many moons. If Neptune has moon orbiting 48,000,000 km from it, then bigger size BD can have influence way farther than this. And celestial body of 450,000 km diameter would have one third the influence of the Sun. Do not think and compare the influence of Jupiter which actual size is only 23,000 km. –All these aspects of the Sun's influence which are almost the same for all big and powerful celestial bodies are something like news for the leading scientists who are stuck with Newton's "gravity" which does not ~~exists~~exist, and all mathematics related to it are fake. So if someone asks the question of who makes the Earth rotate around its axis 24 hours? The answer is the Sun. The moons of all planets and brown dwarfs in our Solar System are PROHIBITED from rotation, who ordered and enforce that? My answer is that all "players" which are the Sun, the planets, and the moons has to perform according to the "script' of the "system." On this the answer is not clear, but here is what I think about it: If a large brown dwarf takes the role of the Sun, the planets outside of the PROHIBITED ZONE would rotate, but if a BD is in the role of a planet, like the 4 "gas giants;" all of its moons regardless of their size are PROHIBITED from rotation. I have stated that the Earth COULD HAVE BEEN MOON TO ANY OF THE GAS GIANTS, and as result of its diminished status would have been prohibited from rotation. Now think about what could happen to the Earth when a stronger celestial body challenges the established order of the Sun, and unfortunately this is happening right now.

DENSITY

Density of the Earth is 5.5 tons per cubic meter and the Moon's density of 3.3. This difference is because the Earth has live BS inside, but the Moon does not. Should our BS density be the same as the Sun's? Kilo per kilo they should be the same. What is certain is that at the initial Big Universal and Local Explosion all black spheres have the same density, but the internal structure might differ. The "material" thrown off from the Big Universal and Local Explosion is homogenous, so what internal rearrangements happen in smaller verses the bigger BSs is to be determined. The scientists' number for Sun's density is 1.4 tons per cubic meter is laughable, when my number is 2,365 tons per cubic meter. Obviously we are quite apart.

Let us concentrate in matter # 1, because matter # 2 is well investigated, but definitely hides more secrets. Matter # 1 cannot exist anywhere except in a black sphere, but matter # 2 cannot exist inside a black sphere – it gets dismantled to BBPA "immediately upon entrance" - how exactly this is done is yet to be determined. I suspect that these outbursts of gamma rays at the poles of a star are indication of this activity, the free energy from this process is expelled.

MATTER AND ENERGY ARE INSEPERABLE

If energy and matter are indestructible, then there is no such thing as "only energy." Then what we call "energy" should be simultaneously energy and matter. We probably do not see the matter, but theoretically they are inseparable. And if they are indestructible, we have to rethink the whole thing. Let us remind ourselves that we are talking about Universe's energy, and not the fossil fuel one.

SUPERNOVA EXPLOSIONS

These supernova explosions are "super killers," and the radius of devastation has to be determined. First is the radius where all planetary systems are completely destroyed. Not only the atmosphere is blown away but the very mantle is thrown out into space. All stars with planets that were the closest to this horrendous explosion would be totally devastated. Let us imagine that such an explosion happened close to our SS. The planets would be torn to pieces, where their cores would be striped from the iron and the ~~silicon~~rocks, and the cores would shine like a mini stars. The Solar System would become flying "debris" of gas, stones, iron, and shiny cores. All this would fly into the inner galactic space. Our scientists see in these devastations little "mini stars" which actually are the cores of planets and brown dwarfs and would call them "scientifically" "protostars" and "star nursery" (nonsense), and the space aliens would wonder what that supposed to mean, but The Little Riding Red Hood would explain it to them. There is one picture called "pillars of star formation" when they are "pillars of planetary devastation." No stars of any kind have been created in our Galaxy in the last 13.8~~7~~ billion years!

EVAPORATION OF BLACK HOLES BY STEPHEN HAWKING

In the early 1970s Hawking "proposed" that numerous mini black "holes" could have been created after the Big Bang. Let us examine his mental ability. He was a good boy - he remembered well what he read, internalizing the information, and had the capacity to explain it back to you – that much about his intelligence – average! What he could not do, is taking a critical look at all the previous physical formulas and mathematical nonsense accumulated before him. To the old scientific ~~garbage~~garbage, he added some of his own. He was approved and endorsed by the central politburo of the ruling scientific mafia that is still in power and distributes scientific nonsense throughout the world. He proposed that after the Big Bang mini black holes were created that could have been as small as the Planck mass: 0.000,000,05. Black Spheres display machine like properties, therefore they could never be as small as sub-atomic particles! The fairy tales continue with Schwarzschild calculations of a radius of black spheres where Einstein's moronic formulas are used.

First question: what the established scientific mafia knows about black spheres, and does it endorse Einstein's formulas? ~~Unfortunately~~Unfortunately, the answer is yes, and if this is the case, they know about black spheres next to nothing. If the speed of light is not the speed limit of the universe, are Einstein's and for that matter any one's else formulas that containing "c" valid? The speed of light is

NOT the speed limit of the universe, therefore wherever you see the use of "c" in any formula – this is a FAKE formula! Neutrinos travel faster than light!!!

"EVIDENCE FOR THE BIG BANG"

Here are 5 evidence of the Big Bang nonsense from "Smithsonian Universe" 2020 page 51. The first one is "BACKGROUND RADIATION." My explanation: The explosion that created the Local Group of galaxies and the Milky Way in it, created this background radiation. The second evidence is "EXPANSION." "If the universe is expanding and cooling, it must once have been much smaller and hotter." I already elaborated quite a lot that the universe is not expanding at all. I do not want to repeat myself, but basically comes down to this - Doppler Shift does not have correct interpretation – it is complete nonsense, and Hubble "constant" for "expanding universe" is baloney – basically it comes down to the WRONG INTERPRETATION of so-called "red shift" which is only caused and directly proportional to the dust and particles the light has to go through. The two teams that concluded in 2002 that the universe is not only expanding, but is expanding even faster, are disgrace to the scientific community.

Here is the next fake "evidence" that they have sucked from their fingers. On p.48 Smithsonian Universe "During the first millionth of the first second after the Big Bang the temperature dropped from 10 with 340 zeroes behind it degrees to mare 10 with 110 zeroes behind it degree Celsius." People of the world, I am questioning the sanity of this figures. Bear with me, I want to write this number in its entirety, it looks to me insane, I refuse to be treated like an idiot by the real idiots. Who came up with this number? Look at their number, and see if this looks possible to you? Here is their initial temperature of the Big Bang:
10,000,000,000,000,000,000,000,000,000,000,000,000,000,000,000,000,000,0
00,000,000,000,000,000,000,000,000,000,000,000,000,000,000,000,000,000,0
00,000,000,000,000,000,000,000,000,000,000,000,000,000,000,000,000,000,0
00,000,000,000,000,000,000,000,000,000,000,000,000,000,000,000,000,000,0
00,000,000,000,000,000,000,000,000,000,000,000,000,000,000,000,000,000,0
00,000,000,000,000,000,000,000,000,000,000,000,000,000,000,000,000,000
degrees Celsius. Quite high temperature, don't you agree? And probably scientists have "duplicated it in lab conditions;" no doubt about it, and I want to see the thermometer that measures this temperature, but I know that the Little Riding Red Hood works for the universe, and I asked her about this huge number, and here what she had to say: "I think that Pinocchio came up with this number, but do not tell him I've told you, because I might lose my job."

Sure enough, the universe has to "cool off" from this insane temperature with 340 zeroes in it, and this is their "definite proof" that there was a Big Bang. If all of us keep silent; the criminals, liars, and fascists are more likely to continue doing what they are doing. In our case the liars will go on lying and keep their cozy jobs and scientific carriers milking the society FOREVER and giving FAKE SCIENCE in return.

The third one is BALANCE OF ELEMENTS. "Big Bang theory exactly predicts the proportion of light elements (hydrogen, helium, and lithium) seen in the universe today." My theory explains way better the existence of the first 30 elements in the universe. In my book is explained the constant release of basic building particles into the interstellar space from stars and the "cores" of planets and moons, and the fact that the easiest self-made elements are hydrogen, helium and so on. I am also pointing out which volcanoes gases have to be analyzed in order to understand which elements are self-produced by the Earth's black sphere as all BSs inside all planets and moons are almost identical in substance but different in size. We have to find out why our Moon's BS died early? My take is that this is an early indication of the end of the life of our spiral galaxy. The thing is that as long as the scientists are looking in the wrong direction, which is the nonsensical Big Bang, the true questions would never be addressed! My theory is fundamental game-changer, and finally the scientists might begin in earnest discovering the REAL universe. The scientists need a THEORY which looks in the right direction, and would lead them to the real discoveries of our larger home – the universe. The scientists at this moment through their lies are leading the human race to its extinction! My bet is that if we do not begin to prepare ourselves to become "space aliens," and if the time comes we would be caught unprepared. As long as the Earth and the Sun function well, we could survive. Could their ignorance combine with rosy scenarios they broadcast might lead the human race to extinction? Be aware that these "Hawking- types" do not know what they are talking about - fundamentally. They do not know the Sun; they do not understand what critical role the magnetic field plays protecting the atmosphere; they are completely unaware that our planet is under "unwelcome" "influence" which could lead not only to bad weather, but to our annihilation; and global warming is caused by the "intruder.!" In the same time, they are busy – doing nothing! Looking to find exoplanets, when do not have the foggiest idea what are the requirements for "living" planet. Pretending that are trying to talk to space aliens, when in reality are doing everything in their power to suppress the messages from them. Is there something more important to watch in the sky, when we have a real emergency in our hands?!

If I have the necessary data I might be able to pin-point with a bigger precision where it is, but even without much scientific data I am saying: Look in direction perpendicular to our North/South magnetic poles, and the "plane" created by the movement of the North Pole. The direction could be determined through what time of the day this move occurs. This disturbance happens twice a day.

Let's look at their forth "evidence" which happens to be GENERAL RELATIVITY. "Einstein's theory predicts that the universe must either be expanding or contracting – it cannot stay the same size." I have read Einstein's brochure about 10 times, I almost memorized it. Surreptitiously, I want to ask some of these "Einstein defenders" have they ever read anything written personally by this idiot? My bet is that they have not. Here are some of his moronic examples: In his writings he uses very difficult language and difficult mathematical formulas in order to confuse and intimidate the reader. Reading his writings at certain point I also felt intimidated, and asked myself can I really be able to investigate this "genius" and mathematician of high caliber? But guess what, not only I deciphered Einstein's nonsense, but also found that his "famous" formula e = mc2 is fake – energy never converts to matter; the time in the universe is one; gravity does not affect the light; and finally both of his theories are 100% garbage. Then reading Newton, I discovered one enormous mathematical error he made about Kepler's third law, which believe it or not, no scientist or mathematician corrected in the last 300 years, and most likely at this moment is in some "supper computers" making "supper" false calculations.

So let me give you some of the nonsensical examples by Mr. Einstein. Here is one: A train is moving with uniform speed. A person inside this train is walking with uniform speed in the same direction which the train is moving. What is the speed (velocity) of this person in relation to the rail track? My guess is that in 1915 the trains in Switzerland were moving with the speed of 30 km/h and the person inside was moving with the speed of 3 km/h. Keep in mind that the accuracy of these numbers is not important. Then he asks~~:~~ ": "You probably will add these two numbers together to get the answer?" My response is~~:~~ ": "Yes, punk, something wrong with that?" After he gives many half-cooked examples like this one, in order to confuse the reader (he never gives the answer right away), finally he gives us his "real" mathematical formula for this task, and guess what is in it? – THE SPEED OF LIGHT! Let me be polite, a question to Einstein~~:~~ ": "F---ing idiot, what the speed of light has to do with the speed of a person walking in a moving train?" In order to compare both answers I calculated it once with just adding the two speeds together (which is the correct answer) and then with Einstein's formula. Turn out that using his formula the answer is a bit smaller. Here how they compare 30 + 3 = 33, when using his ~~formula~~formula, the answer comes down to 32.999. I lost his book and the calculations, but the question stands, what in the world the speed of light has to do with adding two speeds of "smell?" A sportsman bicyclist can travel faster than this train. I wish the idiot is alive and answer this question personally. For those readers that think that I am too obnoxious, I would say this: I consider him a servant of Satan, deliberate destroyer of the science of Astronomy, mathematical crook who is ready to deliver any lie – just to keep his job and image, and in reality he is a mental-retard mistaken as "genius."

Let's look at another of his moronic examples. Let's say that there are two guys, one is in a moving train holding a pebble which is about to be dropped outside from the window. The other guy is standing at the embankment next to this moving train, and will witness the dropping of the pebble. Here what Einstein is claiming these two gentlemen would see: The one inside the train will see the pebble falling "straight" down, but the guy looking at the train will see the pebble "curving" backwards before it hits the ground, and now the "genius" warns us that this would happen only if there is no air present to push the pebble back. Well, I want to tell the genius that his example is moronic where there is air present or not. My version of events is different. Both guys will see the pebble curve before hitting the ground, why? Let's assume that there is no air. Initially the pebble travels with the speed of the train when it is in the hand of the person holding it at the window, but when the pebble is released and goes in a free-fall toward the ground it is subjected to two perpendicular forces: One is the horizontal inertia from the train's speed, and the second is the vertical force of the gravity. The gravity as the perpendicular force would diminish the initial horizontal speed. As the train continues with the same speed, the pebble cannot keep-up with it, so the pebble would be seen as making a curvature backwards before heating the ground by both men. Did Einstein have enough "grey" material in his dumb head to comprehend this? Then why he is explaining something he did not understand, and more importantly, why scientists try to defend his "nonsense" to this day? My guess is that once they accept him as "genius," no one can question his integrity, but I found out that he is an "idiot!" Are we allowed to question his moronic examples now? If Einstein was alive now I would question him why physical experiments should be based on "subjective" information? Did the guy in the train have well~~good~~ enough eyesight and how many beers he might have drunk before this experiment?

A separate book has to be written about the idiot/genius; provided people want it, but sure enough – they need it. His 26 examples are from stupid to more stupid, and I can proof that all of them do not have any merits.

Let's get to the fifth "evidence" for the Big Bang that never happened, labeled DARK NIGHT SKY. "If the universe were both infinitely large and old, Earth would receive light from every part of the night sky and it would look bright. The fact that it is not is called Olbers' paradox. The Big Bang resolves the paradox by proposing that the universe has not always existed." I do not think that this is some sort of a "paradox," and the Big Bang "resolves" anything at all! Light is absorbed by particles and many other obstacles. Contrary to this quote, the universe is infinitely large and old and has existed forever.

We all have been in some high ground in a large city. Imagine that you are in such place in the evening and the street lights are on, and you would be able to see some street lights that are quite far from you. Now imagine that some light rain or fog comes down where you are standing. Wouldn't be reasonable to say that now you won't be able to see the lights that are too far from you? What happened to the light coming from these remote places? It got absorbed by the rain or the fog, because light is energy and also particles. If we are inside a cluster of stars, I bet that the night sky would have been awash with light, but so happens that the Earth is located in pretty quiet and not crowded neighborhood. We see the close-by stars, but they are not too bright. On top of it, we are inside of two cocoons with plenty of gas, dust, and particles present at the edges, and that is the Solar System and the Galaxy. The surrounded galaxies are real far as well. So Olbers' "paradox" is not a paradox at all, but natural phenomenon in which not all light possibly can come to us, because gets absorbed and is blocked by dust, gas, and galaxies.

THE ENERGY IN THE UNIVERSE IS FREE

The whole universe is in a "pool" of energy, and MATTER CONTAINS ENERGY or if we could say MATTER AND ENERGY ARE THE TWO SIDES OF THE SAME COIN. The energy in the universe is never lost in the smallest amount ever, so is the matter. The energy is free because it is part of the universe and never goes to waste, and is inexhaustible and indestructible. Because we do not understand it we assume erroneously that is wasted and one day the universe would run out of energy. This misconception is reinforced through the assumption that matter is converted to energy, and then is lost. When we burn wood, coal, petrol, and gas the energy build in them is released, but this is different energy - chemical. Deceptively this gives the impression that energy comes from matter and is lost forever. That means that one day the universe would run out of energy, and all this is backed by "mathematical gymnastics" and "chewing-gum" physics. And here what I say: You have fooled the entire human race, BUT YOU CANNOT FOOL ME! ... Someone has to move the science of Astronomy forward!

Yes, from the direction of the Sun are coming neutrinos, and mistakenly are taken as an indication of "nuclear reaction" going on inside it, but this "nuclear reaction" is going on OUTSIDE the Sun from the free-association of BBPA. This process is unknown to the scientists. It is widely accepted that the amount of neutrinos "coming from the Sun" is not enough to justify the alleged nuclear reaction supposedly going on inside, when in reality the neutrinos are coming from outside of the Sun.

Looking at the stars emitting all these enormous amounts of heat and light we come to the erroneous conclusion that all this energy is wasted, when it is part of the process in which BBPA become atoms and molecules. The universe is a washed with energy through both the creation and the dismantling of "matter number two." As the Sun produces energy from releasing BBPA, same process is going on inside the Earth, but because the speed of releasing BBPA inside our planet is rather slow it only keeps the interior melted, but allows the formation of crust so we can live on top of it. I am repeating this: We are very lucky that the size of our BS is not bigger than what it is - otherwise the formation of "cool" crust tolerating underground and surface water could have been impossible. All we have to do is compare it to Venus' BS. How surface water could exist there when the temperature is 480 degrees C? By the way, the size of Venus "core" have not been "measured," but I can assure you that it is bigger than the Earth's.

The Sun on its part releases BBPA at slower rate compared to the bigger stars, and as a result of that the Sun "tolerates" Photosphere on its surface, when the bigger more powerful stars cannot "tolerate" photosphere, and the result is that their color is different (brighter).

The scientific community is not aware that the creation of the "matter number two" is nuclear in nature. Keep in mind that for them the universe began its existence with some "Big Bang" BALONEY. The stars for them are like burning matter – mass converts to energy BALONEY, and the "fairy tale" in their heads continues with the assumption that one day the stars will burn themselves out of existence and the sky would be dark, ... but ... what would happen to the Little Riding Red Hood is scientifically unknown – I have tears in my eyes.

In order to prove this in lab conditions somehow BBPA have to be released the way BSs release them, and that is currently impossible. A book is telling me that from the mass of two hydrogen atoms small part of the mass is missing when creating one helium atom. This phenomenon has to be reexamined. Let us have some sense of humor: I am scolding them: "Look for the missing matter until you find it!" In this case we have two hydrogen atoms that are already complete, and energy is necessary for this process for the creation of one helium atom. Around the Sun helium is created from the Sun's wind which consist of basic building particles of the atom and their connecting elements that "are ready and eager" to get together and form the atoms of the helium FROM SCRATCH. When using the word "eager", I am trying to encourage other scientists not to use this trite and basically highly inaccurate Newton's and Einstein's "gravity," but to discover the real nature of things. Once on the internet anonymous person gave me this advice: "It is not important what do you know, it is important who do you know." Sounds like a practical advice for someone's career, but might be a death bell for the real science, and it is! Let's see when the human race will wake-up to the unpleasant reality that the science of Astronomy has become playground for the well-connected, but not for the real and the best scientists. Let the bureaucrats say who can be a scientist regardless of his abilities!

DARK MATTER

Dark matter is a figment of the current "scientific" imagination from the "dark ages" of TODAY! Quote from Smithsonian Universe 2020 page 27: "Evidence that dark matter exists includes the fact that many galaxies would fly apart rather than rotate unless they contain large amounts of unseen matter." There is the logical inconsistency in this sentence: The so-called "evidence" is actually Newton's gravity of falling apples, which has nothing to do with "holding" the stars inside the galaxies. The upper quote could have been as follows: Because we erroneously believe that Newton's gravity holds the stars in the galaxies together, we are afraid that all the stars might fly apart, so we have to "patch" Newton's nonsense with another one, and we came up with some invisible "dark matter." Then another scientist said: "I think that the Little Riding Red Hood and Pinocchio are playing some tricks on us."

PLANETARY ARREANGEMENT IN THE SOLAR SYSTEM

The Sun orbits the CBS, but all the planets and brown dwarfs present in the Solar System orbit only the Sun. The Sun on its part have 9 "predetermined" orbital distances (Newton's gravity has nothing to do with it.). Quite properly the 4 gas giants have taken the outer 4 available places, as if some intelligent being has ordered them there, because of their great power – they need more space. The weakest of them Jupiter took the first of these spots, when the strongest Neptune took the last space. The Solar System, which is a system within the larger system theealled Galaxy, has made this "arrangement." The orbiting distances starting from the Sun to the periphery almost double in size. Despite the current "proper arrangement" that the four "gas giants" have taken to the outer orbital spaces, Jupiter probably "evicted" the "fifth" planet between him and Mars, but could have been an intruder as well. Depends on how strong is the "grip" of the Sun on the planets being in these predetermined orbital spots? And when external celestial body challenges this "orbiting order," could "destruction" of the planet might be possible?

GALAXIES EVOLUTION

Spiral galaxies might not have always have been "spiral." since the BULE. Before they get in this "spiral condition" they might have been looking differently – maybe "lentil?" Seems to me, that the duration of time a galaxy could be in spiral "condition" might be around 5 billion years – this is a wild guess, just to point out toward the direction where the real answers should be sought. Here is my scenario at this moment: Each galaxy goes through three kinds of different appearances, let us say first it is "lentil," definitely the middle appearance is spiral, and the third and final appearance is elliptical, where several (most likely all 40 in the local group) make one "elliptical." After the big explosion all BSs (future stars) might have been "compacted" (matter # 3) and needed time to become normal BSs (matter # 2). This process might have taken billionbillions of years. Life can appear only in the "spiral period." At this moment, our Galaxy is in the last stages of its life as spiral, and unfortunately with the end of being spiral – comes the end of possibility of any life to exist. In case someone wants to go to another nearby galaxy to live, my advice is to leave not only our local group, but the entire cluster or 1,300 galaxies, and find some young cluster where already spiral galaxies exist. Space aliens know more than us, but there are people who always make sure that we do not get their messages through sabotaging our efforts to learn something in that direction. If we pose the question, why should the galaxies go through different forms like elliptical or spherical, the answer is that the universe has ~~goes through~~ a DYNAMIC EXISTANCE WHICH IS PART OF THE "GRAND UNIVERSAL DESIGNE," which is probably designed by no one. There is no such thing as permanency, everything is in a state of constant change, but in a long run nothing really changes. All the changes are the same in this ever-repeating universal cycle. All the "transformations" and "transfigurations" are predictable. ~~, and this change is dictated by this perpetual transformation and transfiguration of the black sphere substance.~~ The black sphere substance "lives" and exists forever in an "inexhaustible pool of energy." Energy and ~~the~~ matter are indestructible and somehow "alive" forever. The universe is a "nuclear reaction" which is in the state of "perpetual motion." The Universal System is in equilibrium – it does not lose or need more energy or matter to function forever.

The closest star to the Solar System is Proxima Centauri which has its own sphere of influence. Beyond that probably brown dwarfs rule the inner galactic space. According to my theory, if these brown dwarfs happens to be the most powerful entities in some "spot;" they are entitled to their own sphere of influence and might have their own "systems" of subordinated smaller celestial bodies, where a ~~certain~~ brown dwarf (BD) is orbiting the CBS like the Sun, and smaller bodies orbit only this particular BD. From our 4 gas giants Neptune probably qualifies as such~~,~~. ~~B~~because one of his moons orbits at the distance of 48,000,000 km, therefore his most remote planet would orbit at the same distance.

SUN'S PROHIBITED ZONE

On its part the Sun does not interfere in the FIELDS OF INFLUENCE of the most planets, except for the first two. The planets do not allow any moons to rotate, and this "no-rotation command" incapacitates them from having their own fields of influence, otherwise we would have "moonlets" going around them. Now, NOT only the planets prohibit the moons from rotating, but the Sun has its SPECIAL ZONE OF PROHIBITED rotation for planets, and in this zone happens to be located Mercury and Venus. Why they do not rotate around their axes? The Sun doesn't allow them. And think about it. Could a moon or moons rotate around Mercury which is too close to the Sun? Also these two planets are real close to each other. Two stars in binary system also face each other for the same reasons. The scientists who do not understand this process in first place have their special terminology ~~like~~of "synchronous rotation." Which could be translated as: "We do not know what we are talking about."

WHERE THE ROCKY MATERIAL ON TOP OF THE BLACK SPHERES INSIDE PLANETS AND MOONS CAME FROM?

All "mantles"~~rocky material~~ in planets and moons has almost the same internal structure. Those that have active volcanoes at this time – have live BSs working at their centers. These mantles are created through accumulation of "matter # 2" on top of the BSs. Our planet has working BS inside which diameter is 1,220 km, but the whole Earth is 12,756 km, this ratio is close of 1 to 10, and that might give us some clue what the ratio of the size of the BS is compared to the rocky planet or moon already created after 13.8 billion years. May we assume that our Moon's BS was one tenth of its size? The Moon diameter is 3,476 km, therefore the size of its BS have been 340 km. Caution, this calculation is valid only for rocky planets and moons with smaller BS than the Earth's. This ratio is invalid for Io and Venus which have bigger BSs than the Earth. Both of them qualify as brown dwarfs, and because have bigger and more active BSs than the Earth's; have a high rate of volcanic activity.

My question is where all this metal and rocky material around the BSs came from? In order to answer this ~~question~~question, we have to know when the planets and the moons were created in the Solar System compared with the creation of the Galaxy. Currently the scientists believe that stars and planetary systems are created at all times. For the formation of planets, they conjured something like rocks getting together and forming "planetisimals" which on its turn create planets. No specifications are given where the Little Riding Red Hood and Pinocchio were at that time.

The "dating" of the rocks should be reexamined. According to my model the Galaxy and the "Solar Systems" should have been created at the same time. On this I am 100% sure. But our Galaxy was probably different at the initial formation and when become "spiral" 4.6 billion ears ago the Solar System "emerged" almost fully "developed." Any way this gap between 13.5 billion years from the initial formation of the Galaxy and the 4.6 billion years of Solar System formation has to be explained. My explanation is that our Galaxy initially was "lentil" galaxy, and stay this way for several billion years, and then become "spiral." During this time all BSs which were compact in "matter # 3 become "matter # 1," because #3 is dysfunctional; stars do not shine, and planets are not formed; BBPA are not released.

After the big explosion the Solar System had all the 200 future BSs "compacted." They should have been 1,000 times smaller. That means that the Earth's BS which now is 1,220 ~~diameter~~diameters; was 1.2 km diameter "compacted" ball of a "white dwarf." I assume that once fully developed all BSs remain the same size to the rest of their lives. Hawking has different opinion, but he knew next to nothing. The "compaction" comes from the explosion. If this happens after supernova explosion; it is only logical that it should happen with the biggest universal and local explosion (BULE) as well, but this is only theoretical speculation at this time. The largest explosion in the universe might be as well some sort of "gentle" explosion that does not damage the BSs, but I doubt that.

From one lentil galaxy becomes one spiral, but the elliptical galaxies are huge and contain the material of several spiral galaxies, may be all 40 of them. In this scenario our Earth might has another 500,000,000 years more to live, provided that our beloved BS or the Sun do not leave us earlier; or some asteroid not hit the Earth; or some supernova does not explode near by; and if we survive the current bad influence.

The rocky material for the mantle came from self-creation by the BS activity rather than from gravitational accumulation from outside. But if some material came from outside; it is subjected to the same heat treatment. It is known that all BSs are releasing BBPA which in their free association are creating atoms and molecules. We also know that small BSs have "tolerance" of "matter number two" to be accumulated on their surfaces. In this case the rocky material in the Solar System cannot be only 4.6 billion years old. It must be billions of years older.

—I have discovered that the biggest "normal" planet or moon is the size of the Earth, but what would be the size of the smallest? The "great" astronomer Hawking is telling us that after the Big Bang there were black "holes" as small as particles of the atom. Of course this is complete nonsense. The black spheres are functioning "machines" that expel BBPA and "dismantle" any atom upon entrance in them, so to say that they could be as small as atomic particles shows complete ignorance! To calculate a black sphere's mass Schwarzschild used Einstein's formula, but let us do not forget that all Einstein's formulas ate fake, as he is one of the greatest mathematical deceivers the world has ever seen. So inside of a BS we have a specific arrangement of the BBPA in such a way that aforementioned functions have to be performed. On TV were shown small stone balls that looks like have been made by black spheres. If we apply the rule of one to ten, then their BSs inside them could have been around 5 centimeters in diameter.

Mars' and our Moon's BS are "dead," our Earth's BS is still functioning. Obviously the life of the BSs is somewhere around 15 billion years. Throughout their lives; once they become mater # 1 from matter # 3; their sizes do not change, but some of them use-up all their internal "material," and on their place inside planets only matter number two remains.

If black spheres do not change their outer size throughout their lives in spiral galaxies, then their outer size is like a carton box. In a sense we do not know is the box full, empty, or almost empty when looking at the Sun, or guessing when the Earth's BS would die. Scientists are calculating how long the Sun would function

53

through some "main sequence" BALONEY. Here is my guessing how we could know how much more life the Sun has. We are close to the end of the life of our ~~G~~galaxy. Definitely when Milky Way collides with Andromeda would be the final blow for any life in the 40 galaxies of out local group. This is the dictate of the Grand Universal Design. For the Sun we have to observe how behave stars like it before they die. Indication for the death of the Earth's BS might be given by the density. The Moon is dead with density of 3.3; when the Earth is alive with density of 5.5. Should the density of the Earth be close to 3.3 when dead? In this case the size of the Earth has to grow quite a lot. It could be calculated, but why bother? The current trouble with the intruder and the displacement of our BS is the most urgent right now.

The accumulation of rocky material and metals from the smallest BS to the size of 1,220 km is directly proportional to the size of the BS. Throughout the life of a BS in a spiral galaxy the ratio between matter # 1 (BS) and matter # 2 (mantle/crust) changes, as through the billions of years through constant release of BBPA the BSs get lighter, when matter number 2 accumulates on top. A lot of gaseous compounds leave the planets. That means that a planet grows in size, as more matter # 2 is created; the BS inside gets less dense, as it does not change its size, constantly pumps out its BBPA; and the overall density of the entire planet gets smaller, because of the loss of the gaseous compounds. The ratio between the sizes of the BS to the outer size of a planet is 1 to n10 at this time, because it changes through these billions of years of existence. We should not forget that "live" BSs in this first PLANETARY SIZE ~~size section~~ of BSs from the smallest to 1,220 km diameter ADD NEW MATTER # 2 ALL THE TIME WHEN FUNCTION ~~add new matter all the time when they function~~. That means that our Earth increases in size without increasing its mass, because we have creation of "matter # 2" (rocks and metal) from "matter # 1" which is BBPA from the BS, but this is simple transfer of matter #1 to matter # 2, where # 1 is 700 times denser than matter # 2 ~~of atoms and molecules~~. The "game" changes when the BSs become bigger than 1,220 km diameter. According to my new and "accurate" classification the celestial bodies with BSs with sizes from 1,220 km to the size of the smallest stars (whatever it is) are classified as BROWN DWARFS (BD). We are lucky that we have exactly the "game-changer" between planets and brown dwarfs, and that is Venus which is simultaneously both. The "brown dwarfs" BSs gradually begin to REJECT matter number two until it is fully rejected with the biggest possible BS for a BD, when bigger BSs than this one are the smallest stars.

In the vicinity of the smaller BSs that create brown dwarfs we see some peculiarities in the sizes of their atmospheres. Of course their atmospheres are poisonous. Io would have had this kind of atmosphere, but it is too close to Jupiter which "vacuums" its atmosphere and leaves it "naked" for our observations. Jupiter's 60,000 km atmosphere "enjoys" the protection of the biggest magnetic field in the Solar System. The bigger the magnetic field is; the bigger atmosphere could be. The Jupiter size without it is around 23,000 km diameter SOLID BODY. The interesting thing is that the Venus' atmosphere which is not protected by a magnetic field "manages" to stays on top of this planet! Such atmosphere without magnetic field has Titan. Looking at this phenomenon, the scientists are coming to the erroneous conclusion that gravity is that holds the atmosphere, therefore our atmosphere is held by gravity, and I ~~vehemently~~ disagree! Valid~~Considering~~ only for "living atmospheres:" "No magnetic field – no living atmosphere is possible!" The magnetic field is the ultimate protector of such atmosphere!!! Today our magnetic field is in trouble, and as a direct result of that our atmosphere is leaking.

HUMANS, WAKE UP TO THE REALITY OF THE SCIENTIFIC MAFIA! At this moment we are in uncharted waters. No historical reference of any kind could give us any clue what the near future could be! Things in the future might not be as dire, but wouldn't you agree that is better to be prepared for something bad that could happen, rather than be caught unprepared? We have accustomed to rely on the Earth to provide us with food, shelter, air to breath, and predictable weather, but things at this moment have changed for the worse, and the most important thing is would it get any worse, and how long would it last? I am saying: Let find out which celestial body is causing this trouble. Then and only then we could have some real projections for the future. Quite appropriately you turn to the specialists for answers, but here I am to tell you that they are 90% brain-washed, and lack the ability to give you any good "scientific" information. Case in point: On the internet I am reading that global warming is 99.9999% human made, but I am 99.9999% sure that global warming, forest fires, unpredictable weather is mainly caused by external influence of another celestial body. Do the scientists know how this is happening? Are you ready to hear the unvarnished truth? They do not have a clue! They are stuck with Newton's gravity which is 99.9999% trash! What are the dangers in this new situation? One of them might be the shortage of food supply; worsening weather; air shortage; burning all forests.

Let us get back to the brown dwarfs' deviation from this rule that from bigger size BS – bigger size of celestial body is created. With brown dwarfs the bigger BSs create and tolerate less of matter # 2, but some abnormally large atmospheres are created around the size of BS that is about less than 10 Earths which is inside Jupiter. In this case we have very deceptively looking "Jupiters." What is characteristic about them are these two features: ~~Firstly~~Firstly, they are capable of having enormous magnetic fields which foster enormous size atmospheres. This is very deceptive feature, where the outer size no longer correlates with power, and the same outer size corresponds to different strength of these brown dwarfs, as the strength of celestial body should be ~~is not~~ determined by the size of its BS ~~that includes the atmosphere but the size of its BS.~~, and not the size of its atmosphere. The scientists in their confusing and erroneous belief that "the stars are filled with hydrogen gas," COUNT JUPITER'S ATMOSPHERE AS PART OF THE PLANET," which is NONSENSE. Brown dwarfs with the size of Jupiter could have strength of up to 130 more powerful than actual Jupiter. By the way they reported of seeing "exoplanet" (their vocabulary) with the size of Jupiter THAT IS WEAKER THAN IT.

The smallest possible brown dwarf is Venus. The next "stronger" but "smaller" brown dwarf is the Jupiter's moon Io. Io is definitely stronger than the Earth – has a bigger BS. Another interesting aspect about brown dwarfs is when we compare Jupiter and Saturn. Observing these two we see that Jupiter is bigger, but the reality is that Saturn is stronger. Jupiter has one of the biggest magnetic fields a brown dwarf could have, and Saturn almost has none. My interpretation of this phenomenon is that Saturn's BS no longer tolerates iron, and without it so is the magnetic field.

TROUBLE INSIDE THE EARTH

First we have to distinguish between the Earth's axis of rotation from the magnetic axis of the south and north poles. So far the axis of rotation is somewhat stable, but the axis formed by the magnetic poles is not. So the switching of the poles is not rotational but is only magnetic. Let us hope that stays that way, because the fact is that rotational axis is around 23 degrees inclined, and I would say that this also has been caused by outside interference as we have right now. I would claim that originally when the SS was created 13.8 billion years ago the rotational axes of all planets were straight perpendicular to the plane of rotation, but some outside influences through these billion of years were responsible for these angles of rotation; and the most dramatic of all Uranus' is outright 90 degree incline. In general; any changes are bad for us. What is perfect cannot be improved. We were just fine the way we were, but here it is; out of nowhere comes some external influence and threatens us with starvation and other bad things.

Why the Earth wobbles at this time? There might be two reasons: The first is because the BS that is off of the center of the planet is 700 times denser than the mantle. The second could be the direct influence of this "intruder." Could the current episode of external influence move this axis of rotation as well? That would be very dangerous if done suddenly and substantially.

Mars continues to rotate around its axis, but its BS is "dead." This means that the rotation of the planets has nothing to do with the BS inside whether it is "live" or "dead." My knowledge of electric motors is limited, but here is the conclusion I have reached. The Sun makes the molten iron together with the rocky material rotate around the planetary axis, and about this I am 100% sure. Some scientists think that the BS inside rotates even faster, but I think that BSs do not rotate at all, and this is also in the "orders" of the Sun. The created friction between stationary BS and the rotating iron creates the magnetic field, and all this comes from the "System" enforced and carried out by the Sun.

The switching of the magnetic poles will not kill us, but the displacement of the BS could. It is off of its base some 400 to 500 km. In order to create proper magnetic field has to be right at the center of the Earth. Everything around it is melted due to the high temperature. The melted metal around it is probably thinner in the direction of the dislocation, which happens to be toward Indonesia. These big earthquakes in the Indian Ocean and the recent one off the coast of Japan might begin to make sense. They are caused by the extra heat under the crust. A high volcanic activity is naturally to be expected, and eruptions in New Zealand and Philippines have happened. Indonesia is volcanically active area, and I expect some volcanic activity there as well. More volcanic activity and earthquakes in this part of the world are to be expected.

If this is a celestial body that is causing our trouble, it is below the Earth's equator, and about 20 degrees above the planetary plain. It has an angle from the plane of planetary orbiting~~rotation~~ determined by the line of movement from Canada to Siberia. Its relative speed also can be calculated as its orbit is close to elliptical around the Sun. Judging from the severity of its influence on the planet Earth; this thing is getting increasingly closer to us~~, and is very powerful and dangerous~~. Reportedly the movement of the North Pole from 70th parallel in Canada toward Siberia has increased in speed in the last 10 to 20 years. Through this increase we can envision the speed with which this thing is moving, assuming that this intruder is located perpendicular to the magnetic North/South axis. The worst influence might be yet to come, and might be at 33 to 45 degrees on its trajectory above the Sun, and then we have to endure the same bad influence on the other side. These projections are made on the assumption that the proposed trajectory is right above the planetary orbital plane.

Keep in mind that Mars' BS was not able to return back to the center of the planet 700 million years ago. This is scary, if out BS would not be able to return to the center, we are doomed, but it has returned before. Reportedly the rate of losing atmosphere is not "alarming" according to the leading scientists, but how bad it could get, and how long it would last? Keep in mind that no one knows why this is happening before this publication. If this loss of atmosphere continues for a prolong time (a speculation – 200 years), or the rate of loss is increased, then Tibet and Nepal might find themselves in difficult-to-breath environment. The celestial body causing this has to be tracked down and better predictions based on true data have to be made.

I can assure you that all planets that have angles in their rotations are caused by "outside" interference of black spheres, stars, brown dwarfs, white dwarfs, and neutron stars, and have not been in any "collisions." And let us not forget that stars and brown dwarfs are black spheres, and their different names come only from their different appearances, when their internal substance and structure are the same, but the most dangerous and difficult to detect are the white dwarfs and powerful neutron stars, and their structure is different – they consist of the most ~~possible~~ compacted matter in the universe.

The Earth's poles have been reversed before; Uranus remained with one of its poles pointing toward the Sun. That means that after the outside influence on the Solar System (SS) was gone; some planets remain in some precarious positions and cannot return to normal. Wikipedia is reporting that some pattern is seen in these reversals every 200 million years. Let us not forget that this is the time that takes for our Sun to make one complete circle. Could that suppose to mean that the SS has entered in some dangerous zone? Plus, according to my theory "everything" in our Galaxy has to orbit around the central BS; then we have to come to the conclusion that some "foreign" to our Galaxy celestial bodies might have come from outside. I do not know where these "Magellanic Clouds" are located. On top of it they are reporting that these pole reversals could take from one lifetime to several thousand years. The faster movement of the North Pole speaks for acceleration in this influence, and I would say increased intensity.

Our BS is 400 to 500 km off of its center. Would it return back to normal when the intruder is once and for all out of SS? I think that it would return to normal, as long as it does not go any farther. Should the rotation of the mantle push it back to the center? I would say, yes.

Looking at its capability to influence all planets and peripheral brown dwarfs (Uranus and Neptune) I would say that if the intruder is a close by, but scientists cannot see it. If it is a brown dwarf it should be around 300,000 km diameter located in the Oort cloud, but if it is some weird thing like these little but extremely powerful white dwarfs, then it ~~should~~might be under our noses.

What is the danger to our lives and the Earth from this intruder? The negative consequences have been enumerated, what our concern should be is how close to the Earth it would get (the closer, the worse), and how long it would be around (the longer; the more damage to the Earth; and the more suffering for us). We have to develop a new "detector" for these "fields of influence," but in order for the scientists to do that, they have to put in the trash one of their favorite nonsense called Newton's gravity and Einstein's "scientific" TRASH! All black spheres (CBS in the center of the Galaxy, stars, brown dwarfs, and planets) are in mutual communications, because they are in the "system," and to have more info how this system works, we have to break the code of their "communications."

THE DIFFERENCE IN ROCKY MATERIAL AND IRON
IN DIFFERENT PLANETS AND MOONS

The live BSs inside planets and moons release BBPA and new "matter # 2" is created. Then the question could be: what is the rate of this "creation and accumulation?" In "Smithsonian Universe" 2020 expressed the idea that enormous amount of material is "produced" by the Io's volcanoes as lava flies some 300 km above the surface. ~~I would caution them that this material coming~~ The material coming out of these volcanoes is not all "new," this is melted rock which is in the way of the gases that have to get speedily to the surface. Courteous observation of our Earth does not suggest of some rapid creation and accumulation of new material is happening, but I strongly believe that such growth is happening right now through creation of new matter in the mantle of the Earth, and the newly created matter is pushed out through the volcanoes, or lava is coming out of so-called "shield volcanoes," or ~~fishers~~ on the bottom of the ocean. Still, it seems to me that planets and moons that have live BSs inside right now for the work of more than 4.6 billion years have not accumulated substantial amount of matter, and if we look at Io, I would venture to say that it never would get any bigger than what it is, because it is a brown dwarf. Only BSs smaller than 1,220 km diameter ACCUMULATE MATTER THROUGHOUT THEIR EXISTANCE.

Here is how I see the difference in accumulation rocky mat~~t~~erial and iron inside of the planets: Once again reminder that there is no structural difference between planets and moons, so when we talk about planets; the same is valid for moons. My contention is that smaller black spheres generate smaller heat than the larger ~~once~~. If this is the case, then here how the rocky material would differ. Let us start with comparison between our Moon and the Earth. The rocky material from the Moon would contain some chemical compounds that are missing in the Earth rocky material, because inside of the Earth they have been evaporated due to the higher temperature. Venus' rocky material should lack some chemical compounds compared to the Earth's because its BS is bigger and generates a higher temperature. Io should lack more chemical compounds found in the Earth and Venus, as the temperature rises. All of them produce iron. Looks like the biggest producer of iron is Jupiter, but Saturn which is more powerful than Jupiter, no longer tolerates iron and probably does not have any rocky material as well.

In "Smithsonian Universe" 2020 they are saying that between Mars and Jupiter should have been a planet four times the size of the Earth. Yes, there should have been a planet there, but I do not know where "Smithsonian" got this "four times bigger than the Earth?" There are no "planets" bigger than the Earth; according to my classification! Advice to the reader: Do not believe any of the nonsense from Newton, Einstein, Schwarzschild, Eddington, Doppler, Hubble, and Hawking.

CREATION OF PLANETS AND MOONS

~~The small BSs are visible from the "footage" in the sky coming from supernova explosions where the melted mantles of many planets are blown off, and their BSs shine like mini stars. These are the "live" BSs like ours. How the Earth and the Moon were created? At formation~~ In the SS planets and moons were created together 13.8 billion years ago; they were two BSs that happen to be next to each other. One was bigger and created the Earth, and the smaller created the Moon. As

59

the bigger of the two BSs, the Earth took the "orders" from the Sun, because this is in the "scripture" of the Grand Universal Design. The order from the Sun was simple: ": "Orbit around me at this 'third' available orbital spot at 1 AU; completing it in one year" according to Kepler's and Bode's laws. Once again, the orbital distances and speeds are predetermined by the Sun. Could we say determined by the size of the Sun?

The fact that the most powerful brown dwarf in the Solar System Neptune has taken the widest space available which is the last one; and all four large BDs have taken the outside (wider orbits); and The the 4 small rocky planets have taken the 4 inner spaces; plus Jupiter (the weakest of the 4 BDs) has taken the narrowest space of the last 4 spaces speaks unequivocally that this is intelligent arrangement, but ;is but is done by the Sun.? Probably it is in the Grand Universal Design prescribed for the "Solar Systems." That means that when the SS was forming; the planets and the brown dwarfs were shoveled around by the Sun until they got their "prescribed orbits, and the rule was "the bigger; the more peripheral orbital spot should take." Between Uranus' and Neptune's orbital space is almost half of the entire orbital distance of all planets and BDs in the SS.

What should be done with Newton's gravity? What shall we do with Mr. Newton' "gravity?" I am the investigator, prosecutor, jury, and the judge, and the verdict is in: ": "Mr. Newton you have been found guilty as charged." Newton's gravity is valid only for falling objects to the surface ground on any celestial bodies y. But wait a minute, every dog and monkey knows that, if something is thrown up eventually would go down; then did Mr. Newton discover anything? In the final analysis he "discovered" nothing! And I suspect that iron should fall faster if there is a magnetic pull by the Earth. For this experiment two balls of equal size and weight should be made, but where one should be from contain iron and the other not. The iron ball should hit the ground first, or if they are in a small plastic mesh thrown from airplane with recording device or observed by a parachutist, the iron one should pull down the sack down. Parachutists, please do this experiment for the sake of science!

LOOKING FOR EXOPLANETS

It is good to look for exoplanets, as long as the "lookers" know what to look for, but that is not the case. Live atmosphere and existence of animal and plants could happen only on a rocky planet which is not bigger than the Earth, because the Earth is the biggest planet in the universe that can have life and living atmosphere. When I say "the biggest," to the size of 12,756 km; it could be 13,000 or 13,500 km, but no more than that! I guess that in the big distances they are looking; such small bodies cannot be seen. What they see are brown dwarfs, and no life is ever possible on top of them! But I read that they believe that it this is possible! How

60

they can measure the surface temperature, and decide the size of the BS inside, and is it working and at the c~~ea~~nter of the planet? Impossible!

Because without working magnetic field; no living atmosphere is possible; and consequently no surface water is possible. This leads to the conclusion that pure drinking water on the surface could exist only when is protected by living atmosphere with the ultimate protector of both – working magnetic field and right size of BS. Magnetic field can ~~exists~~exist only if the BS inside is exactly at the center of the planet. Planets orbiting binary star systems cannot have their BSs at the center~~;~~, therefore no living atmosphere is possible.

What are the so-called "gas giants?" The scientists have labeled them planets, and I disagree. My definition of planet is the possibility of spacecraft to land on it and an astronaut to walk on it. No one can land on these "gas giants" – they are hybrids between planets and stars. If we call the "rocky planets" "planets", then these supper-radioactive all-metal-melting hells where the entire surface is like live volcano should have a different name. With their larger BSs tolerance for "matter # 2" existing on top of them diminishes, so if Jupiter has some rocky su~~rface~~bstance, it probably is looking like boiling thick kasha of magma. On Saturn which has bigger BS probably there is no rocks even melted – <ins>almost</ins> all metals and rocks have been evaporated by a higher temperature. Visits are not recommended; observe from afar! Starting with Venus the surface temperature begins to get higher and higher until <ins>a</ins><ins>the biggest BS of BD become</ins> star~~s is born~~. As I understand some Soviet spacecraft landed on Venus, but eventually malfunction<ins>ed</ins> due to this high temperature. The scientists looking for exolanets are seeing BDs, therefore it is out of the question that they can see live planet.

I have been thinking about how the space aliens travel these real long distances, and have come to these two conclusions: First of all they travel with UFOs. We do not know what speeds ~~this vehicles~~this vehicle can achieve in space. Second, the distances are enormous – what about their exercise, food, and fuel. The fuel should be atomic. For exercise, they probably hop from planet to planet to stretch themselves, and that is how they replenish their food and fuel. Still the distances are in ~~thousand~~thousands of light years. I suspect that they might be traveling with speeds exceeding the speed of light. One thing is for sure, they do not travel with rockets. Do they have children trough their journey so that they continue the journey that lasts ~~thousand~~thousands of years? How children could be raised in a spacecraft? Yogi could stop their hearts and hibernate.

PROPOSAL FOR OPENING COMMUNICATIONAL INTERNET CENTRE WITH THE SPACE ALIENS WITH ME AS A BOSS

Space aliens can tell us a lot! But where is the "Center" for gathering such information? Space aliens do talk to humans, but for security reasons the messages are suppressed. Without free exchange of information, research in this area is difficult to impossible. I can understand the intelligence agencies, but let me be clear: The best minds in the world need access to this info! This is only possible in a free exchange of info, and my guess is that from the best and free minds of the world the security agencies can get far more! Because the SURVIVAL OF THE HUMAN RACE IS OF THE MOST IMPORTANT CONSIDERATION IN THIS EFFORT! Plus, there are people that make things up – crop figures, for example. I propose opening of international organization for open communication with the space aliens with me being the boss. The gathered information would be available to the entire world, and I would make sure that is not like Wikipedia where everyone can say what the wants, and any one can erase what they want! Let us do it, and do it NOW!

Right now we can and we should visit only the Moon, and that should be out first destination. Plus, if we can learn how to survive on such uninhabitable place would be a step in the right direction. We should learn how to survive in any celestial body. We should keep in mind that there are two kinds of planets and moons in our Galaxy, and the difference between them is that some are "alive" and some are "dead." These that are alive and have a working black spheres; have the potential for "warm temperature" close to the volcanic vents, even the outside temperature might be minus 200 C. Such places in the SS are the moons of the gas giants with live volcanoes but not Io (too powerful). On one of these moons that are "alive" we can count not only on heat coming from inside which we can use for some food production if we carry seeds, but also we can count on continuous "production" of chemical elements and useful compounds like water and others. Our Moon is "dead," that means that its BS has stop working millions of years ago, and all we can count on there are the existing rocks, but we should learn how to live in such environments as well. Water on the Moon could be available, because when a BS dies the underground water as little as it might behave been goes toward the center and percolates toward the surface when gets in contact with "supposedly" hotter center. On the Moon we can gain knowledge about the procreation in this low-gravity environment. By the way, no moon in any "Solar System" is "allowed" to have a magnetic field, but that is a needed protection from the "solar" wind. So, living atmosphere on any moon in the universe cannot occur naturally, so why not learning how to create partial one when we definitely would need it. This is one of the things we can develop and test on the Moon.

What should be the size of the future astronauts? I saw a woman in Los Angeles very skinny less than one meter tall. I think that astronauts should be small skinny people, and the reason for this is that they would consume less food, because on the Moon cannot be raised that much food, and on their journey hopping from one dead moon to the next, food might be a serious problem.

Now I want to mention one inaccuracy of the current "scientific fog": The Moon and the Earth supposedly are binary system – sounds like baloney. We have seen how important it is for our BS to be at the center of the planet. I challenge the notion that the Earth and the Moon are orbiting the Sun in some sort of "binary system." The Moon MUST orbit the Earth – that is how the "system" dictates.

Let's begin establishing what the small BSs do. First of all, how small a BS could be? According to mentally-deficient Hawking right after the Big Bang there were some very small BSs, so small that were like sub-atomic particles, and you know that these kind of scientific nonsense can come from him and from his hero - the mental-retard Albert Einstein. Any way, a BS in order to function under no circumstances can be sub-atomic, because it has some mechanism in it for pumping out these BBPA. On TV was showing some small spherical stones, and I thought that these might have been made from some of the smallest BSs. In the name of science one of these balls has to be cut in half and analyzed, so that we know what becomes from nonfunctional BS, and this probably has been done already. There is a question what should be the ratio between BSs of planet/moon relationship? Earth's BS is 1,220 km, and our Moon's BS when was "alive" probably was 340 km diameter, so this ratio is about 4 to 1. Another interesting question would be: Can the Earth be a moon to Io with reported BS size of 1,800 km diameter, if Io was where the Earth is, and the Earth next to it? Definitely Io would have bump out of this orbiting space the Earth, and definitely no live atmosphere would have developed ever on top of it. With this ratio of 1,220/ 1,800 km I do not think that the Earth could become Io's moon. Io would have developed "atmosphere" similar to Venus's without life or liquid water. Let us say that the Earth have bumped out Mars. Then maybe the Earth would have Mars' fate and die 700 million years ago.

It is interesting that in each orbital spot around the Sun there is only one planet, but in the 5th orbital spot their millions of rocks. Could that mean that each orbital spot has the power to handle many stones, or only one planet or a BD, but cannot handle two or more planets?

EVOLUTION OF GALAXIES, STARS, AND PLANETS

There is evolution, but is different from what the scientists think it is. For them some "evolution" is going on throughout the entire universe since some Big Bang that created it all. We know that this is a fantasy and a wrong approach that could lead only to nonsensical conclusions.

Let the truth be told. There is a constant evolution throughout the entire universe, and in the same time there is no evolution in long run what-so-ever. Whatever evolution is happening; firstly, it is confined only in the Grand Universal Cycle; second it is always the same; and third beyond the confines of the Cycle there is no evolution in a long run. Nothing stays the same – everything evolves, and is very simple, well organized, and easy to understand. Let us say that all galaxies in a cluster are the same age. We know that they would go through a (approximately) 20 billon year cycle, and this cycle repeats absolutely the same over and over – forever. So where is the evolution? The evolution is in this period of 20 billion years, and because all cycles are absolutely the same, once you learn one – y, then you'll will kknow all of them all. On the sky we could watch different stages of the same cycle in different locations.

HOW SOME OTHER BLACK SPHERE CAN INTERFERE WITH THE EARTH WITH NEGATIVE CONSEQUENCES FOR THE LIFE HERE, AS IT IS HAPPENING RIGHT NOW?

As we already know all stars are almost "naked" black spheres – our Sun is not quite "naked", because as a "weaker" star it allows formation of photosphere on its surface where bigger and more powerful stars have stronger "wind" of BBPA, and photosphere do not have a chance of formation. Brown dwarfs are smaller than our Sun, but develop something like "fog" on top of them just enough so we cannot see their shining light coming from them. We could call them mini stars, because that is what they are – they are pure black spheres just with a smaller size. And because they are weaker; they allow accumulation of some matter # 2 on top of them. For the most powerful of them this cover is only gases.

What ever the intruder is, we have interesting phenomena going on right now. The "intruder" influences differently our BS (matter # 1) than the rest of the entire planet (matter # 2). And, think about ithat; the Sun makes matter # 2 to rotate around its axis; then most definitely the intruder "wants" Earth's matter # 2 to rotate differently and orbit differently. The Sun has predetermined orbital spaces; so does the intruder, but "his" orbital spaces are pointing in different direction. As a result of these conflicting orders our BS is off the center. The intruder moves the magnetic poles to be in perpendicular position from its equator. I already spoke about the "communications" among the BSs of which the scientists do not know anything about, but it is about time to address this issue again.

One conclusion can be made for sure – BSs threat differently "matter #1" than "matter # 2." Obviously we have a tug of war between the Sun's "field of influence" and the "intruder's." For sure they are perpendicular judging from Neptune's rings. Whether the Sun has better grip on the "matter # 1" (BS) or "matter # 2" is unknown, but for sure this displacement of our BS points to different "treatment" of these two matters. The Sun has a firm grip on the entire Earth, but the "intruder" successfully "manages" to misplace our BS. , hHow exactly this phenomenon takes place? If I have access to scientific data, I might be able to answer this question. This displacement is serious matter that has to be addressed with the attention it deserves – our lives depend on it. If there is a bigger displacement of our BS, the magnetic field would get proportionally "weaker."

which immediately would lead to faster rate of atmosphere loss, increase in solar radiation, and consequently more forest fires and unpredictable weather. Yes, the "intruder" can kill us all if it gets too close to the Earth and nothing we can do about it, but these "disturbances" had happen many times before, and life continues. Time to act is now - this is an emergency. All efforts should be concentrated on finding the "culprit" in order to predict accurately its future trajectory, and know with certainty should it come any closer, and how long would be around. Reliable forecast is needed; otherwise humanity would not know what is happening – the same way the dinosaurs did not know.

Bode's and Kepler's laws are real laws of the universe. They are "enforced" by the Sun. Bode and Kepler was the real scientists – they observed and try to make sense of it. Today scientists are brain-washed from schools and universities. I wanted to verify something on the internet-sewer, and was promptly informed that what I am interested in is given to 10 graders – therefore I am stupid and lacking education. Immediately on the right side of the screen appeared some weird offensive "recommendations!?."

Using words about the Sun like "field if influence" and "orders" are necessary first to make the reader aware that Newton's gravity has absolutely nothing to do with planets orbiting the Sun or moons orbiting planets. Let us determine how far the gravity of our planet really extends? Pieces of broken satellites fall to the Earth's surface, therefore the "gravity" do extends there, but what makes the Moon going round the Earth is NOT the aforementioned "gravity" – it is a different force. If the scientists do not even "dare" to imagine this "reality," and stubbornly continue to "believe" that Newton is correct, no progress in this direction could be made. The Moon orbits the Earth because the size of Earth's BS has this power to "dictate" this orbiting. Who knows how many moons Jupiter's BS has the power to "command," but sure enough there are 79 of them.

At this very moment we have some sort of emergency that the scientists do not have a clue what is it? Firstly, they do not know about the existence of black sphere inside of the Earth or any planet, brown dwarf, or that all stars are 100% black spheres. Second they do not understand these "fields of influence." What the intruder's influence consists of is that it "tries" to "order" planets in the Solar System to orbit around it instead around the Sun, which enters in contradiction of the established order created by the Sun. This is the cause of the so-called "iron core" displacement. We have a chaos in the entire Solar System right now! Who knows what influence it inserts to all moons right now? All planets and moons probably have some leeway in their orbits. As I said before, our Moon has no right to get farther from the Earth 3.3 centimeters each year, and I stand by this statement, but now is not the time to verify this, because we have an external "body" that is doing weird things to all planets and moons, and is not out of the possibility that the Moon might move a bit.

Although we know that the BS is a sphere - it has a definite polarity – being a magnet. That is one of the clues we have for its internal structure. We can never be inside of black sphere, so we have to gather all of the external "evidence" in order to gain knowledge for its internal structure, therefore we have to relay on observations and speculations. We can say with certainty that in the equatorial plane circular orbital channels do exist. Could we imagine them like hollow hoops? My bet is that any matter # 2 (rocks and metal) once entering in this "hoop" which is wide for the Earth let us say 15,000 km, and for Neptune maybe 70, 000 km, might have "hard time" exiting. Also if BS enters might have even harder time exiting. Fact is that there are millions of rocks in this "vacant" orbital distance between Mars and Jupiter. How they gather there. Here is my bet, they enter and could not get away. Some of these asteroids move real fast, then how strong is the power of this orbital "hoop," because they could not continue their journey? That means that these hoops a very dangerous, and as the intruder moves through the SS and somehow the Earth enters in one of its hoops, then our trouble might increase substantially. The speed of the orbiting planets diminishes from center to periphery, but the size and the strength of these celestial bodies' increases. The question is can we create a model for all sizes of celestial bodies? If we had this chart available right now, we could have predicted the size of this intruder, but if it is white dwarf which I highly suspect, the size might be 1000 times smaller.

THE HISTORY OF THE SOLAR SYSTEM BEFORE THIS BOOK

Quotations are from "Smithsonian Universe" 2020 because it reflects the current ignorance of all scientists in this field throughout the world. Quote from Smithsonian Universe 2020: "The Solar System is thought to have begun forming about 4.6 billions years ago from a gigantic cloud of gas and dust, called the solar nebular. This cloud contained several times the mass of the present-day Sun." Analysis of these two sentences: Firstly, I challenge the timing of formation 4.6 billion years ago, must have been way earlier. Today scientists perceive the creation of stars and planets in "clouds of gas and dust" using the word "nursery." The reality is that this gas and dust are created by the stars in first place, and all of them happen to be black spheres, and the notion that stars and planets are created at all times in some "nursery" is complete and absolute nonsense! These "nursery" are left over debris after supernova explosions. The smaller shiny "proto-stars" are BSs of planets and BDs. For the creation of a "Solar System" as the say is happening on the site of their creation are necessary many small BSs. Without their presence these bodies cannot

66

be created, and their presence is possible ONLY after one of the Big Universal and Local Explosions.

And of course the scientists do not know that when the Solar System was created except the Sun; another more than 200 black spheres were needed for 9 planets and 140 moons and spherical rock, and untold amount of very small BSs for real small rocks. All these requirements point only to one conclusion – the Solar System and the Galaxy were created at the same time!

The quotation continues: "Over millions of years, it collapsed into a flat, spinning disk." Why the scientists use the word "collapse?" Once again the culprit is Newton's non-existing gravity! The "field of influence" is where the equator is, there is no need of anything to "collapse!?" This field of influence is rather obvious for the Central Black Sphere located at the center of our Galaxy, and that is the reason why the form of it is like disc. Part of the Sun's field of influence is the plane where all planets orbit, and the same configuration is for all planets that have many moons. I do not see any reason for some "collapsing!?" The text explaining the formation of the Solar System reads like moronic explanations that suppose to be science, but sound more like badly written fairy tales. "FORMATION OF THE PROTOSUN:" "Under the influence of gravity, the solar nebula condensed into a dense central region (the protosun) and diffuse outer region (the protoplanetary disk)." Notice again that proverbial "gravity" is doing its miraculous "job." ~~These people do not know what they are talking about, but sure enough they know how to take the MONEY appropriated for this field; write books; being on TV; being professors – liars for money in the field of science – cancer to the scientific thought – starting with Newton, Einstein, Hawking, and today's "do-not-knows". The system of scientific deception is on!~~

"3 RINGS AND PLANETICIMALS": "Instabilities in the rotating disk caused regions within it to condense into rings under the influence of gravity. Very gradually, planetisimals (small objects made of rock or rock and ice) formed in this rings." "The planetisimals attracted each other by gravity and collided to form planets." Notice this absolute nonsense – "planets made of rocks and ICE. Help, planets are being made from ice!!! Absolute insult on human intelligence, and again there is no specification where was the Little Riding Red Hood at that time, and these rocks, aren't they wonderful, beating each other to such high temperatures and creating planets. On TV they are showing them like pieces of burning charcoal. Quote from "Universe" by Smithsonian p. 101 "Many of the planetisimals came together to form Moon-sized bodies called planetary embryos, which finally underwent series of dramatic collisions to form the rocky inner planets and the cores of the of the outer gas-giant planets." ~~I asked one cow what she thinks about this, and here is the answer was: "mooohaa" – cannot be translated.~~ Maybe you have noticed these new words like "protosun", "planetisimals", and "planetary embryos." In one textbook for college students I counted roughly 800 new words in Astronomy coming from nonsensical formulas and theories, designed to torture the young generation of students to memorize them when giving them no REAL knowledge!

KONSTANTIN POPOV'S UNIVESAL LAWS

#1. THE UNIVERSE IS PERPETUAL "NUCLEAR REACTION." MATTER AND ENERGY ARE FOREVER "ALIVE" BECAUSE THEY ARE INSEPARABLE AND "LIVE" FOREVER.

#~~1~~2. ~~.~~ The Universe has existed forever, and will exist forever, because BBPA live forever. They are "alive," because they are constantly unstable, and require the Grand Universal and Local Explosion to come to some temporary "equilibrium."

~~#2. In the constantly repeating Grand Universal Cycle no BBPA are ever lost or any of their energy.~~

#3. BBPA are not convertible to energy and vice versa.

#4. Gravity is one of the many "influences" any celestial body has, and does not extend beyond Roche's limit. The overall influence extends without any gravity toward other celestial bodies. Stars stay in galaxies; planet orbit stars; and moons orbit planets because of the "influence" between them and have nothing to do with regular gravity.

#5. The Universe is one giant system with smaller systems in it. All~~The larger~~ systems are identical. In these systems are smaller and smaller systems one inside the other. In a "super cluster" there is a "cluster" like the "local group;" inside them are galaxies; in them are "solar systems"; and in them there are planetary systems. This is not a complete list.

#6. All these systems go through the same "evolutions," which are part of the Grand Universal Cycle which is always the same and last~~ing~~ approximately 20 billion years ~~separately for each super cluster~~.

#7. Basically the Universe has only three kinds of matter: "matter #1," "matter #2," and "matter #3." All BS and stars consist of "matter #1." White dwarfs, and neutron stars are "compacted matter #1," and that makes them "matter #3." All compound particles, atoms, and molecules are "matter #2," and their ~~which~~ existence is only temporary.

#8. All rocky planets and moons are the same (from the new classification); except Io which is a brown dwarf (BD).

#9. All "live" planets, moons, and brown dwarfs have both matters #1 and #2. Matter #1 is inside their BSs which is 100% the same ~~like~~ in ~~-~~stars. Matter #2 is all metals~~;~~, rocks ~~the mantles~~;, the crusts~~;~~, and the atmospheres. When BSs inside them "die;" like in Mars; the whole planet consists only of "matter #2."

#10. Current believe that stars and planets are created at all times, even now, are illusions. Stars ~~They~~ are created only at the beginning of the "C~~e~~ycle~~;~~." 13.8 billion years ago (for the Milky Way). The planets and the moons were created later from the released BBPA. ~~and for our cycle it was 13.8 billion years ago.~~

#11. The Universe is not expanding or contracting at all! It is in everlasting "stationary" equilibrium, but goes through constant "transformation" and "transfiguration" confine into the Cycle.

#12. The Universe is ~~G~~going through never-ending repetition of exactly the same Grand Universal Cycles, where we have the appearances of "aging" and "death~~,~~" but in reality the BBPA ~~actually~~ never lose any amount of matter or energy, so all these "evolutions" are one mimicry which is part of the Grand Universal Design.

#13. The Universe is in the state of Perpetual Motion going through apparent "transformation" and "transfiguration;" creating the appearance that is "alive" and "lives" forever. The only thing that lives forever are the BBPA and their energy. Atoms are temporary formations; with ~~a temporary appearances~~ a temporary appearance.

#14. The Universe is never-ending nuclear reaction in equilibrium. Nuclear reaction UNKNOWN to the scientists!

#15. The Universe is in an inexhaustible pool of energy that is forever ~~preserved and~~ present and preserved.

#16. There is no energy without matter, so the light is matter. ~~#16. The galaxies, stars, and planets are temporary formations. They go through apparent "evolutions," but it is all from the script of the Grand Universal Design; through repeating endlessly Grand Universal Cycle; it is like watching exactly the same "movie" over and over again.~~

#17. There is not a single hydrogen atom in any star anywhere in the Universe.

#18. There is only one Universe in all directions and it is not spherical, neither "flat."

#19. Living atmosphere is protected exclusively by ~~the~~ a magnetic field. No magnetic field; no living atmosphere!

#20. Magnetic field is created through the rotation of the metal from matter # 2 around live BS, which has to be in the center of the planet. If the BS is not at the center of the celestial body regardless of the rotation, there would not be a good magnetic field that can protect the atmosphere. ~~This rotation is done by the Sun in the Solar System.~~

#21. All BSs inside planets and smaller BDs like Jupiter create iron, which is ~~after~~ subjected to double gravity (regular and magnetic); finds itself right on the top of the BS. Saturn does not have much iron~~,~~ because has ejected it.

#22. BSs inside rocky planets can be easily moved from the center of the planets. Mars lost its atmosphere 700 million years ago, when its BS was yanked from the center of the planet ~~out~~ – all the way ~~to~~ under the ~~its~~ crust. Currently Earth's BS is 400 km off of the center of the planet, and the ~~direct~~ result ~~that~~ is "bad magnetic field," which leads to a leaking atmosphere, global warming, forest fires, and bad weather.

#23. Earth's BS is fine when "governed" only by the Sun's "influence," but when external "influence" challenges and interferes with Sun's "influence;" removal of our BS from its base happens. The consequences of that are not good. With a small influence; we get small displacement, but if the influence gets stronger the result might be catastrophic.

#24. All chemical elements from the Periodic Table are produced by the Earth's BS right here inside the mantle, including gold and heavy elements.

#25~4~. The survival of the fittest might be for the animals, but the survival in the Universe is for the knowledgeable.

#26~5~. In other "solar systems" in our Galaxy, the central star or the brown dwarf that commands any particular system arranges the subordinating planets according to their power. The most powerful to be at~in~ the periphery, and the weakest should take the closest to the "Sun." ~inside orbital "lots."~

#27~6~. ~In s~Supernova explosions are not from inside a star but from outside, where hydrogen explodes; as a result the ~"exploding"~ star gets~is~ "compacted" to the max, and becomes neutron star or white dwarf. ~T~There cannot be more compacted matter than this one. (Just a speculation)In this explosion probably "small pieces" of it get~this compacted matter are~ separated from the "main body~.~" and become white dwarfs. ~They immediately take spherical shapes, and are extremely powerful, but look like mini planets, when packing 1,000 more power!~

#28~7~. A JOKE: To the three dimensions of x y z Einstein added "t" which stands for time, and that is how all clocks in the Universe got broken or inaccurate. I am adding fifth dimension "i." Now we have five-dimensional field xyzti, and 'i' stands for idiots like Einstein.

#29~8~. A JOKE: "The gravity of a black hole is so strong; that even light cannot escape." I think that it is even stronger – and sucked-up the brains from Einstein's dumb head, Eddington, Hawking, and many others.

FORMATION OF MILKY WAY AND THE SOLAR SYSTEM

(Some recapping and new thoughts.)

The Solar System (SS) was~is~ created together with the initial formation of the Milky Way. After one of the Big Universal and Local Explosion happened, it created the local group of galaxies to which the Milky Way belongs. All the stars in our Galaxy were present at the initial formation, and not coming from outside or being "created" in some "dust and gas clouds" later on. The Central Black Sphere instantaneously took "command and control" with its "field of influence" in disk-shaped form. All smaller black spheres that happen to be in this disc of influence of the CBS are under ~her~its "command and control," and let us assume that this disc is about 100,000 light years across.

Here what the scientists are saying: "Stars tend to congregated in galaxies." This implies that stars are wandering through the universe and for some reason "decided" to "congregate" in galaxies (maybe for a cup of coffee), holding to each other by guess what - Newton's idiotic gravity!

Here's how galaxies in a cluster are formed. All galaxies in a particular cluster are the same age, and are going together through the evolution in the proverbial Grand Universal and Local Cycle, and all this is coordinated by the Grand Universal Design. Otherwise if every small group of galaxies goes through separate Big Explosion, then there would be constant super strong galactic explosions in the middle of a cluster, which will have very disruptive effect on the "development" and evolution of the surrounding group of galaxies. The formation of galaxies in a cluster begins with series of BIG LOCAL AND UNIVERSAL EXPLOSIONS. The biggest black spheres in the universe which contain the matter from 40 to 50 galaxies are exploding in relative short time or maybe simultaneously. From the mass of the previous 50 galaxies - 50 new are created. The number 50 is not that important. The important thing is that the mass from the previous 50 galaxies is accounted to the last sub-atomic particle. Compare this explanation to theirs, where the whole universe was packed in something as big as the size of a walnut! Help, can some one come-up with more moronic idea than this one?

LET ME TALK ABOUT THINGS I AM NOT SURE OF. The "matter" here on Earth incorporates energy from the Sun, and this process looks to me is predominantly happening only in environments with "living atmosphere" like here on Earth. All plants and animals are made from this matter consisting of atoms arranged in chemical compounds, which foster creation of life, and contain "chemical energy." The chemical compounds have energy incorporated in them, and this is the energy we take from the fossil fuels, and the food that we eat. What Einstein's "famous" formula has to do with this process? - Absolutely nothing! When we burn a log of wood in the fireplace in order to get the heat out of it - in the process the log is "lost," but that is because this log contains simultaneously matter and chemical energy. Let us do not forget; matter in the level of BBPA is indestructible. The same energy that comes from the Sun, comes from any celestial body with a "live" BS inside; namely from volcanoes. That means that underground in some "live" planets or moons some form of primitive life could be found. Matter and energy are always together; therefore, there is no energy only without matter with it.

Water is available throughout the universe because is easily self-produced. The scientific investigation of what elements and chemical compounds are self-produced by a functioning BS is very easy. It could begin with analysis of whatever gases are coming from universal volcanoes, and let us not forget that here on Earth there are "subduction" volcanoes as well. At this moment when the process of "self-production" of atoms and molecules has been just discovered, there is no way that things could be categorized with some complete accuracy. The new thing is the knowledge that plasma, atoms, and chemical compounds are produced throughout the universe in spiral galaxies, and what is coming out from the Earth's "universal" volcanoes is approximately the same stuff coming from any volcano in our Galaxy. Different galaxies go through different evolutions, but all is in the Cycle, predictable, and the same!

The BSs come in different sizes; creating different celestial bodies. The bigger and more powerful of them dictate to the smaller once what to do. For sure the most widely used order is: "Orbit around me!" I read that scientists observed that a white dwarf destroyed the planets of some other "Solar System." My interpretation of this is that, if another Sun comes in the Oort Cloud; our SS would be destroyed. The power of the Sun over the subordinating planets is DEVASTATINGLY STRONG. IF ANOTHER CELESTIAL BODY SERIOUSLY CHALLENGES THE SUN'S ORDERS, PLANETS WOULD BE DESTROYED IN THIS TUG OF WAR. We are told that there is some "angular momentum" and if you apply some force on a celestial body it would move in some other direction - nonsense coming from Isaac Newton. Things do not work that way! The Sun or its size makes all planets orbit at prescribed distances and speeds. On top of it makes them rotate around their axes when prohibit rotation of the BS inside. The report that White Dwarf has destroyed planets in a planetary system means that the grip of our Sun on the orbiting planets is so strong that the Earth cannot escape Sun's influence in one piece, but it would be destroyed!

Right now our Galaxy takes part in the "local crunch," and together with the galaxies in our local group are heading toward the so-called "Big Attractor." How at the end of the universal cycle," and end of the local "crunch" the galaxies would "know" where to congregate for the next big explosion? There must be some (I call it) "Grand Universal Design" (GUD)," which we must decipher thoroughly. The galaxies might have some build-in memory in them in order to "know" what to do and where to go. These two Magellanic Clouds are not here close by to our Galaxy by chance; this is in the script of GUD. The script might be that first, the small galaxies have to gather around the big ones Milky Way and Andromeda; and second the two groups have to collide. All galaxies in the local group are getting together to form one huge elliptical galaxy in which all "matter # 2" would be "dismantled." Then this elliptical galaxy would get together with other elliptical galaxies and form a quasar. In this case the One of the Biggest Black Spheres (OBBS) would create maybe from 100 to 200 galaxies after the Big Universal and Local Explosion.

The fate of our Galaxy and all stars and planets in it is dictated by the Grand Universal Cycle. The Sun is not allowed to go anywhere on its own. Some times the scientists are showing arms of galaxies, as if they are "going" somewhere on their own, when this is a temporary illusion, because they are not "allowed" going anywhere.

The scientists are telling you that the Sun has 6 billion more years of life. The Sun will die when it gets to the end of its BBPA supply, but they are not aware of this. They live in a world where the stars are filled with hydrogen gas, but this is not the reality. The Sun's fait is dictated by the Grand Universal Design. The current "pole-reversal" probably has nothing to do with ultimate end of "matter # 2" existence, which might be coming in the next 500 million years. The "universal cycle" dictates what should happen, and is underpinned by the DYNAMIC existence of the black sphere substance. What we observe in the sky is this constant TRANSFORMATION AND TRANSFIGURATION of "matter # 1" and "matter # 2" caught in the Grand Universal Design. Different clusters of galaxies are in different stages of one and the same Universal Cycle, but at different stages.

The question why this Cycle has to be repeated forever remains, but might come down to this intrinsic DYNAMIC EXISTANCE. Somehow the entire Universe has entered in this "never ending" repetition of one and the same Cycles, and because the energy is FREE, and matter ultimately is never lost; it is in perpetual equilibrium. It would be reasonable question: If the universe is a "system," then why our Earth is in trouble right now? People were assuming that God created the universe, but that is not the case. There is God, which teaches us how to live and behave, but the Universe is independent entity and is not DESIGNED TO SERVE US. We have to adapt or perish, and I am trying through telling you the truth; that you would know what measures to take in order to continue the existence of the human race. Personally I am pessimist, but God ask me to teach you, and I have to obey. I think that God has plan for you, which is to SURVIVE – take the knowledge and survive! My job is not over; I have to complete the investigation of the science of Astronomy; and train young people. At this moment humans even remotely cannot adapt for the simple reason that their current knowledge about our "bigger home" which is the Universe are 90%85% WRONG. Virgin Mary spoke to the three children shepherds in Portugal: Cesar N. draw the Bible with empty pages: Maya Indians predicted with their calendar that in year 2012 we are to enter in a new era, but who was to listen and "interpret" these messages from the space aliens? Maybe we could survive, but only if we take the appropriate measures. Please understand, in the previous human history there was never such thing happening before. What might happen, depend on severity, the human race might be confronted with prolong food shortages. If you take the appropriate measures, you might survive, if not, all bets are off. Problem is that there are all kinds of crazy people talking all kinds of trash, but I would say, use your head; the GOOD sometimes wins over the Evil. I think that already people have enough EVIDENCE that might be pointing in the direction toward the HOLY SCIENTIFIC TRUTH, AMEN!

Let us assume that the space aliens know far more than us, and it is obvious that they do, but if I am correct, the conclusion that no living creatures might continue to exist between cycles is inevitable. In order to survive they have to transport themselves out of a cluster of galaxies, but the distances are so enormous that makes this endeavor impossible at any speed. The way I describe the universe, the cyclical of creation of life and it's the most advanced mental development cannot survive from Cycle to Cycle. Our fate is to be recycled: "See you after 9 billion years or so into the new 20 billion year 20-billion-year cycle, where we have to go through the same evolution from fish to rats to current idiots and so on. By the way, I read in Wikipedia that Indians have this knowledge of "9 billion years" of "rest" or "death." Pray to the Earth to be good to us; the Sun to shine; and the intruder to go away - if you have nothing else to do. Remember, currently "Mother

Earth" is "sick," because of some outside influence. It does not mean that this is the end of the world and our existence. I think I know how to find the intruder and predict with ~~e~~some sufficient accuracy what is in store for us, but it takes a village, and the~~if no~~ scientific community has to believe me~~takes me in; I'll be dead in the water~~. Second thing is this monitoring of the North and South magnetic poles. If this is not done; no scientific data might be possible~~achieved~~.

WHO MAKES THE PLANETS TO ROTATE?
(Recap and new thoughts)

Are the planets rotate around its axes on its own or the Sun makes them do it? The answer is that the Sun makes them do it. Now we know that some planets have live black spheres inside like the Earth and some like Mars do not. I was puzzled that even as Mars after it lost its BS still continues to rotate around its axes with absolute precision and reliability of 24.63 h. Obviously the "death" of its BS did not affect its rotation. What that tells you? And something else, MARS WOBBLE SOME 670 MILLION YEARS AGO ~~REAL~~REALly BAD! From 700 million years ago to 30 million years ago. Its BS which was 700 times denser than the rest of the planet and 10 times smaller diameter of 680 km (Mars' diameter 6780 km). Today Mars does not wobble, because the rotation is done by the Sun, and as the extra density disappears where the BS was located.

All asteroids in the asteroid belt rotate, and that phenomenon indicates that the Sun makes them rotate~~;~~, therefore it rotates all planets as well. And quite puzzling, it rotates even planet that have an angle of rotation. Let us look at some "rocks" in the asteroid belt: 951 Gaspra 18 km. long rock – rotation period – 7.04 hours; 2867 Steins 6.67 km. diameter – rotation period 6.07 hours; 21 Lutetia 121 km. diameter – 8.17 hours; 243 Ida 60 km. long rock – rotation period 4.63 hours, and there are more examples. All these asteroids are from the Main Belt between Mars and Jupiter. If there was a planet in this "Belt" as supposed to be, who would have made this planet rotate, its BS or the Sun? And now we have the answer – it is the Sun, otherwise this "rocks" would not rotate, and that is a part of the UNIVERSAL COMMANDING INFLUENCE carried on by the Sun. There is something else going on as well. Looks to me that the black spheres inside planets and brown dwarfs are "prohibited" by the Sun from rotation, and that is the reason for existence of magnetic fields. But if the BS is displaced from the center it ROTATES WITH THE REST OF THE PLANET! Our BS stays STUCK toward Indonesia. Why the BS and the rest of the planet do not move together intact? Obviously the "rotational grip" of the Sun is so strong, that even the BS could be "caught" in this rotating matter # 2. Looking for possible explanation; has to be postponed. Here what experiment I suggest to the space agencies. Go into the asteroid belt; let the spacecraft stay there still; I bet that the Sun would make it rotate around some axis and orbit the Sun. This spacecraft would orbit the Sun with the approximate speed that all rocks. If the spacecraft decided to exit this "orbital spot" EXPECT "difficulties;" the Sun do not allow EASY EXIT; an effort would be necessary. This "effort" has to be recorded for future analysis.

ORBITING AND ROTATIONAL INFUENCE OF THE CELESTIAL BODIES

~~Finally~~Finally, we have to bury Newton's gravity. Planets do not move at constant speed, and there is no such thing as "angular momentum." The Sun together with the entire Solar System orbits the Central Black Sphere (CBS) located at the center of the Galaxy, because the CBS makes them do so. All stars in the SS are under the CBS orders to orbit around it, and this is because of its influence. This influence is equally strong throughout its field of influence. There are rather big stars at the very edge. The Sun has similar influence. It has this field around its equator where all planets orbit the Sun. It also has this spherical influence that encompasses the entire SS. Similar spherical influence has the CBS, but on smaller scale; but its equatorial field is enormous. We seek absolute similarities in all sizes of BSs, but we are not going to find it. I am examining all these influences in order to find what and how external celestial body affects the Earth. We have to explain the displacement of the BS of the Earth! Let us not forget that planets and brown dwarfs in the SS do not allow rotation of their moons. I would say that this no rotation is ordered by the system which is the SS, rather than each planet enforcing its rules. The Earth could have been moon, and then would have been prohibited from rotation. No planet in the SS has such enormous size moon compared to its size. I wonder does the Earth have the extra power to handle another moon, and could it be that our Moon demands all the strength from the Earth's influence?

Let us assume that a large celestial body has uniform spherical influence, and let us call it the weakest. The field around its equator where the moons are orbiting is stronger, but the strongest influence is in these orbital distances. So let us imagine that the alleged intruder affects negatively the Earth at this time with its weakest influence which is the spherical, but if one of his predetermined orbital distances crosses the Earth's path, then the damage might be more severe. This might answer the question: Why 700 million years ago our neighbor Mars get severely affected, yet nothing ~~happen~~happens to the Earth?

LAGRANGIAN POINTS

Lagrangian points are the most obvious in the orbit of Jupiter. According to a textbook these Lagrangian points are caused by the gravitation of the Sun and Jupiter. I disagree; they are caused by the repulsive "element" of the overall influence of Jupiter and the Sun's "orbital tubing." It has been established already that all celestial bodies posses a repulsive part of their overall influence. We know that the Sun does not pull objects toward it as Newton claims, but pulls them to go around it. So this is one aspect of the Sun's influence that scientists do not interpret properly. In this case I have to describe it thoroughly. In case of Jupiter, we have rubble in his assigned orbital spot. Celestial bodies do not want to be hit by other objects. These "rubble" does not have any substantial speed toward Jupiter; this repulsive aspect is not that strong; but strong enough to "stop" these stones. So Lagrangian points are not at all caused by the gravity, but quite opposite, they show the repulsive nature in this particular feature of the overall influence of Jupiter one side, and they capturing strength of the Sun's "orbital spot." These stones have entered into Jupiter's orbital spot, and found themselves ordered by the Sun to stay in this orbital distance. At the end of previous century, we observed a predicted collision where Jupiter was hit by some asteroids. Somebody might say: "Where was the repulsive aspect of Jupiter's influence back then?" The answer is that this repulsive aspect of the overall influence has certain strength, and it was not sufficient in this case. In the Lagrangian points it works, but in the case of these asteroids the repulsive force was overwhelmed.

HUMANS AND OUR PLACE IN THE UNIVERSE

There is one song that says that we are "dust in the wind." I would modify this: We are live creatures on top of some melted rocky material on top of a BS until things are favorable for our precarious existence. God is mainly for our morality, and in that we failed as well. Whatever Gods there are, and whatever their powers are; the universe is supreme and obeys only its own rules. The Earth is part if it, and part of the universal "cycle." Because our BS is currently attacked; we have weather trouble, forest fires, global worming, and atmosphere loss. That would lead many people to question their belief systems. Allow me a joke: "Join your local fascist party that pretends to be a party of GOODNES. Go to church and pray to God to save you from Satan, and continue serving the one with the horns." The current trouble with our planet will test our morality, philosophy, religiosity, you name it. The most important word in the future world might be FOOD. Somebody might have gold, but can you eat gold? Gold can buy you food, if food is available. So here is my advice to governments right now: Save food for at least 3 to 30 years, and find a way to preserve it and store it properly – sounds rather insane, but you have to do it in order to survive. This process takes years in time, effort, dedication, and taxes. In 2008 banks in US run out of money, the solution was easy – give them money, but when you are confronted with lack of food – where could you get it? Let us look at insurance industry. It is based on money, but if the world is suddenly confronted with total lack of food supply, the insurance can be only in food that has been properly and safely stored.

77

INFLUENCE OF ALL CELESTIAL BODIES IN SPIRAL GALAXIES

First of ~~all~~all, the bad influence of another celestial body is based on power and size except white dwarfs. I would be talking of the influence of another Sun, but let this be understood~~:~~. Our Sun can be compacted to the size of the Earth without light, and even if a fraction if this size gets closer to the planetary system the results are bad. The influence of celestial bodies have ~~been established~~been established, but through repeating it I want us to see what would happen when in an already established system like SS enters "uninvited" powerful celestial body. Let us say that a powerful but invisible white dwarf enters the SS. Its influence is "saying" to the planets: "Orbit around me!" But this influence enters in contradiction to the established influence of the Sun. Our planet together with the rest of the planets and moons in the SS get this "contradictory orders." What I am trying to establish is what is happening to the Earth at this moment, and what "eventually" "might" happen in the future. Let us look at the "orders" of the Sun, so we can understand what would happen if another strong celestial body~~"Sun"~~ enters into the SS. Keep in mind that the scientists do not have clear understanding of the "influential power" of the stars, the brown dwarfs, and ~~especially~~ white dwarfs. They still live in prehistoric nonsense like gravity, when there is no such thing. For some reasons of coincidence the Newton's WRONG ~~moronic~~ formulas do "work" and sure enough this does not help the situation. To establish the influence of another "Sun" as an intruder in the SS is not as simple as the ~~moronic~~ notion of gravity; it is ~~way~~ more complicated. For example, the Sun has predetermined orbital distances, and in the asteroid belt millions of rocks are "captured" and "ordered" not only to "stay" in this orbital distance, but also to rotate around some "artificial" axes! Stones with long shapes do not have "axes;" then there is only one conclusion – they are artificially created by the Sun. As a result of this powerful outside influence~~intrusion~~ we have the core of the Earth displaced. We need a real scientific explanation immediately, or we do not know how to protect the human race from extinction. The scientists are not asleep at the wheel – they are simply ~~100% missing~~85% brainwashed. Talking about complicated influence of celestial bodies! So here comes our intruder with its influence and orders all "matter # 2" to rotate around some new direction, when creating its own orbital spaces that are "moving" with it. It is obvious that the BS (matter # 1) is treated differently than the mantle "matter # 2." We have to determine where from this "new influence" is coming, but no one understands "celestial body's influences" yet. I have a solution to this problem, and that is the constant monitoring of the magnetic poles of the Earth on both sides. The eventual disturbances would point toward the intruder.

The influence of celestial bodies comes from their mass, and that is how the white dwarfs have 1,000 stronger influences than stars. All stars are pure black spheres, and do not change their size from birth to death, except when affected by supernova explosion. This outside explosion not only "compacts" the star that supposedly explodes, but MAYBE tears up pieces of it currently called "white dwarfs." or using some other confusing and exotic names.

Let us look first at the influence of the Central Black Sphere that commands the Milky Way. At the equator 50,000 light ~~years~~years' radius all celestial bodies have to obey its orders. If there is a group of stars in some spot, the order goes to the biggest one. If in some spot the biggest celestial body happens to be brown dwarf, the order goes to it. The order is simple: "Orbit around me with this prescribed speed." The CBS does not interfere into the influence of this celestial body, and that is how the planets in the Solar System do not orbit directly CBS but orbit the Sun. The Sun has its "independent" sphere of influence, but all this is part of the "smaller system within a larger one~~the system~~." If another CB~~V~~S appear at the edge of the Milky Way and has the same strength, then all "solar systems" in that half of our Galaxy would be negatively affected, and not only the planets but the "path of the "suns" as well. The Sun determines the orbital spaces of all planets. This determination is expressed by Bode's law. Scientists do not accept this law because of "mathematically deceptive principles." Second thing that the Sun determines is the speed of each planet expressed through Kepler's third law, which Newton "improved" with negative consequences. Newton and Einstein were "great" mathematicians. Today the great and highly respected MATHEMATICAL AND SUPER COMPUTERS/ARTIFICIAL INTELLIGENCE STUPIDITY REIGNS!

WHAT IS A PLANET?

Here is my definition for a planet~~:~~ "~~:~~ "Planet is a celestial body that a space craft can land on it, and astronaut can walk on it," from the 4 rocky planets these are: Mercury, Earth, and Mars. Venus has 4~~6~~80 degrees C.; how could anyone walk on it? The 4 gas giants are absolute hells. There is no difference between moons and planets. According to this definition, each rocky planet could~~our~~ have been a moon. Let us say that at this orbital spot where the Earth is right now, at the formation of SS was only our Moon. In this case our Moon would have rotated around its axis, and definitely had some thin atmosphere.~~Moon could have been a planet. If the Moon was on the place of the Earth, it would had atmosphere and life, but this atmosphere would have been thinner, and maybe could not have supported large animals and humans.~~ In my classification Jupiter, Saturn, Uranus, and Neptune are brown dwarfs, because they are too powerful to be planets, and the name "brown dwarfs" "stays" with them until the largest of them qualify as stars.

Here I have to make a short explanation of the word "law": Bode's and Kepler's laws are real, because they are the "laws" of the real "dictate" of the Sun. Their discoveries are "real," when Edwin Hubble's law of expansion of the universe is baloney, because the "Expanding Universe" is a fiction. The 4 gas-giants are brown dwarfs (BD) together with Venus which is exactly on the border line between planet and BD (astronaut cannot set foot on it – it is too hot), plus, believe it or not, we have a moon that qualifies as a BD, which is Io – one of the Jupiter's moons. So we have the terminology "planetary system," but if we have to be sticklers the verbiage should be "system for planets and brown dwarfs." I know some things in this book are repeated, but because of their novelty – it has to be adequately explained.

Let me go over the size of a black spheres and what they will become: smallest (size 1) – spherical rocks; a bit bigger (size 2) – moons and planets; still bigger (size 3) – brown dwarfs: bigger than that (size 4) – stars; the biggest in each galaxy Central Black Sphere (size 5), it is located in the center of all galaxies and "commands" it all; and finally the biggest ever BS which I would call "One of the Biggest Black Spheres" (OBBS) containing the mass of 200 ~~50~~ galaxies or so (size 6). When OBBS explodes we have one of many explosions that are the biggest in the universe, but they might be very gentle so that BSs are not damaged because~~as~~ they are "machines." So these "explosions" might not fit our precondition notion of "explosion" – it might turn out to be some electromagnetic fast separation, let us call it conditionally a "special" ~~yet~~ explosions. But maybe after these explosions what is flying around are "compacted" BSs which are (matter # 3) neutron stars and white dwarfs. ~~Why don't call it a "special explosion?"~~

Between our characterizations of planets, brown dwarfs, are stars are some "hybrids" when switching from one category to the next. Something interesting has to be mentioned about appearances, especially now when we have to find the intruder which could be a brown or white dwarf, and interferes with the Solar System's planets, and the most concerning - our beloved BS. Its size should be no less than 300,000 km diameter if brown dwarf ~~if it is at the edge of the SS which is about 1 light-year away, and could Eris if it is a white dwarf~~, and if white dwarf –ONLY 300 KM DIAMETER! ~~Let us assume that Eris is a white dwarf; then its power is 1,000 times stronger than Pluto which is almost the same.~~ To complicate the issue farther; scientists invent new names, when fundamentally there are only 3 categories of matter already described. ~~– # 2 (atoms and molecules), # 1(BS), and # 3 ("compacted").~~

~~Jupiter looks big, but most of its substance is gas, when melted rock, metal, and its BS are only 23,000 km diameter. If its size of 142,900 km was solid, then it would have been 130 times stronger.~~

We have galaxies that look differently, but all of them MUST have central black spheres; scientists say: "some" of them!~~;~~ All stars are pure black spheres, and all planets, moons, and "spherical stones" are made initially by BS inside them. All of them have "influence" which is proportional to amount of mass they have, and approximate correlation to outer size which has a deceptive quality. I am reading that the scientists are using Kepler's third law to estimate the amount of matter from the Solar System to the center of our Galaxy. I ~~, and~~ wonder how they can assume that his "law" would apply to the Galaxy when it applies to the Solar

System ONLY! This "law" is based on the assumption where the distance between the Sun and the Earth is to be taken as 1.0 (1AU), and the time for one orbit it taken as 1.0 (1 year) as well. My question is: When applying this law to the central black sphere, what takes the place of the Earth? If there is no satisfactory answer to this question, then all mathematics is wrong, and I am sure of that! Although the central black sphere and the Sun are made of the same substance, it does not mean that their "influences" would be the same, and they are not. The stars in a galaxy are not orbiting the central BS the same way planets are orbiting the Sun, and moons are orbiting planets. The similarities are obvious, but mathematical nonsense is unwelcome! Tomorrow, when the scientists build the "artificial" intelligence, things would get even foggier. Imagine a scientists standing in front impressive-looking computer, and by all means with some human-looking head with a tape-recorder inside it, and the scientist is saying: "Oh, you, supper intelligent computer, smarter than all of us combined, please, tell us what the nature of the universe is?" And the tape-recorder might respond when simultaneous moving the rubber lips and the eyelids are blinking: "Ask Konstantin Popov – he will tell you, because he knows better than me, and I am only as smart as the nonsense they have cram in me."

MERCURY

All 8 planets have predetermined orbital distances. They are "arranged" by their "sizes" to certain extend. The smallest planet Mercury took the first available orbital distance from the Sun; when the strongest brown dwarf was placed on the last and widest orbital space. Titius discover these numbers, and Bode in 1781 promoted this "law." Here are the numbers for the first four planets: Mercury $(0 + 4) / 10 = 0.4$; Venus $(3 + 4) / 10 = 0.7$; Earth $(6 + 4) / 10 = 1.0$; Mars $(12 + 4) / 10 = 1.6$. The distances from the Sun for all regular planets calculated through Kepler's and Bode's laws and the actual measurements almost coincide. When I use the expression "commanding" influence, I encourage other scientists to get to the bottom of these "commands". Compare this with the moronic expressions like, "bending the fabric of space." Bending what? How much is it bended? What is this fabric of the space? No one asked these questions the certified moron Einstein. ~~Do you believe that the space has a fabric, and of W~~what material this fabric might be?

~~When I say that the Newton's gravity is only for "falling apples" I mean it. A piece of iron with equal weight and equal size will fall faster to the ground because is subjected to double force: The first one is the regular Newton's gravity and the second one is electromagnetic force. Experiments should be conducted.~~ Back to Newton's fake gravity - in order for it to work for the Sun, the planets have to move with some "constant" yet "predetermined" speeds in a straight line into the "preordained" orbital distances in the "plain of the orbiting field~~planets~~." The planets should have "known" these distances and speeds prior to the time of being "captured" by the Sun's "gravity." Sounds like a scientific "manure!" Have the scientists ever seen some planets moving like this, and being "captured" by some stars?

~~Talking about Sun's INFLUENCE, N~~now we know that each orbital "~~spot~~distance" a planet occupies is PREDETERMINED. Next peculiar "order" by the Sun is the "no rotation around its axis", and applies for the first two planets when enforced by the Sun, and most likely through its size. This "no rotation zone extends ~~more than~~ 108 million km from the Sun which is the orbital distance of Venus. The Earth at 150 million km is made to rotate by the Sun. Now you know why the first two planets do not rotate – they are not "allowed" by the Sun's "commanding influence". According to Kepler's third law the radius of orbiting around the Sun cubed should equal the time for a complete orbit squared, but there is very important stipulation – for all planets these two measurements should be determined only in comparison to the Earth's measurements which should be taken as 1.0 for the radius and 1.0 for the time of one full orbit around the Sun. So the radius of the Earth's orbit is 150 million km., and should be taken as unit of one (1.0). The time of one complete rotation around the Sun is one year, and no one should be allowed to use days, hours, or whatever units the year consists of, or the Kepler's third law would be INVALIDATED. Guess what? This gross mistake was made by the "great" mathematician Isaac Newton! And after him for 300 years all mathematicians and scientists did not see this absolute mistake and correct it!? Even though we are smarter than monkeys, there is no specification how "smarter" we have to get in order to discover the universe. Using this "wrong" formula - the Sun's mass has been ABSOLUTELY MISCALCULATED! And then we have the nonsense of Eddington's assumption that the Sun is filled with hydrogen and some nuclear reaction is going on in the middle of it - you will realize that today's scientists know about the Sun next to NOTHING! Yes, they have satellites watching it, but cannot see much, and even closing their eyes to the reality that the Sun has a surface, but this contradicts the notion that is "filled with gas!" This is gross! Looking at the surface of the Sun, and not seeing it? And these people called themselves "scientists?"

Mercury has the most elliptical orbit, why? Here is my explanation: Contrary to Newton's no existing gravity, according to which the Sun suppose to pull objects toward itself, and Einstein's moronic idea of "bending the fabric of space" - the Sun has a REPULSIVE FORCE which gets stronger closer to the Sun, and celestial bodies that are moving closer and closer to it are repulsed and "DIRECTED" to go around the Sun. So the Sun pulls asteroids and comets not toward itself but channels them to go around, so the Sun do not really pulls matter toward itself, but its "commanding influence" DICTATES their movement in predetermined orbiting channels to go AROUND. The scientists are talking about "angular momentum" which is ~~probably~~ BALONEY. Very interesting aspect of this is that their speeds are neither increased nor decreased. The Sun does not "like" things to fall on its surface, so is Jupiter - that is why we have the so-called "Lagrangian points." The "repulsive force" of black spheres is valid only ~~for any~~ to the larger size celestial bodies – beginning at 1,220 km diameter of BS, or current outer size planet of 12,756 km.

The Sun is orbiting the Central Black Sphere with 220 km/sec, with the same speed is moving also its "field of influence," which is the entire Solar System, and that is the reason why planets have elliptical orbits. WHY MERCURY HAS THE MOST ELLIPTICAL ORBIT? BECAUSE IT'S LOCATION IS IN THIS REPULSIVE ZONE AND HAS THE HIGHEST SPEED OF ALL PLANETS! Let us examine Pluto's orbit. It has an angle of 17 degrees toward the field of the majority of orbiting planets in the Solar System. Pluto's elliptical orbit is very revealing about the Sun's "field of influence." The next conclusion could be made for all black spheres' "fields of influence:" Firstly we can say that all of them in spiral galaxies have this "equatorial rings" that "command" smaller black spheres to orbit them at particular distances and speeds. Judging from our Sun, we can say that these orbital rings get progressively larger as the distance from the commanding BS gets bigger and bigger. Second the speeds of the orbiting celestial bodies are predetermined and the highest speeds are next to the "commanding" BS, going slower and slower toward the periphery. ~~Although this rule applies for the Sun, looks like do not apply to the Central Black Sphere commanding our Galaxy. I might be wrong on this one, but at the periphery of the Galaxy stars are moving at way higher speeds than our Sun._ If this is not the case, then these apparent "arms" could not have been formed. Although there is a drag; the peripheral stars cannot keep up with the required speed. If we compare spiral galaxy to a bicycle wheel, and make two markings one in the middle of a spoke and the other at the rim, then when rotating the wheel, the marking at the rim would move faster than the one in the middle.~~

For comes and ~~asteroids~~ asteroids, the repulsion is only to make them not to fall directly on to the Sun but to get slinked back to where they came from, and its remarkable how they are going back exactly where they were before. So this is another aspect of the Sun's "influence field" which on one side has this "plane" of predetermined orbital distances for planets and brown dwarfs with predetermine speeds for orbiting where our Earth is, but if some celestial body is out of this "plane" also falls ~~"enters"~~ into "predetermined pattern to ~~of rotation or~~ going around." This applies for bodies like Plato. ~~Scientists are talking about some sort of "angular momentum" which might be "angular BALONEY!" Help, Newton's gravity is trying to kill me! Let us get back to Plato.~~ For some reason it found itself in this "unusual orbital influence" of the Sun with the current orbit, which is in the Sun's "arsenal of different influences," but the scientists "refuse" to see it of what it is!

Mercury is in a special location that happens to be in the closest "predetermined" orbital distance, and the highest of all planets orbital speed (47 km/sec) yet in the "strongly repulsive and defensive area" of the Sun. The "no rotation order" from the Sun could be explained this way: If Mercury is allowed to rotate around its axis, then it potentially can have moons, and when these moons find themselves between Mercury and the Sun there would be serious "repulsive" problems for them.

VENUS

Venus is the second planet from the Sun. Its predetermined orbital radius is 0.7 which from 1 AU or 150 million km is 108 million km. Smithsonian 2020 page 116 compares Earth and Venus "The two planets are virtually identical in size and composition." May I "inform" them that they are not identical in composition (melted rock in the mantle) and in the size of their black spheres, and even though the size of Venus' "iron core" (BS) is not known, I can assure you that it is bigger than the Earth's. Bigger BS releases BBPA faster, which leads to a higher temperature that "evaporates" relatively more elements and chemical compounds. ~~This difference is theoretical at this moment and has to be ascertained, but I believe to be 100% accurate.~~

~~Firstly~~ Firstly, the size of any BS is what it is from the Big Explosion after it become matter # 2, if the BSs initially were white dwarfs and neutron stars. If they are such, then they need time to recover and become regular matter # 1. All black spheres are matter #1, and are filled with BBPA which they release. The big stars never release all of their BBPA, but the smaller like our Sun release their BBPA until they are "empty." This should be true for all brown dwarfs and planets as well. In our Solar System we do not have "dead" brown dwarf, but we have "dead" moon (ours) and dead planet Mars. Their black spheres have exhausted their BBPA and seize to exist. Their "immortality" consists in the fact that their BBPA have turned themselves to "matter # 2" creating celestial bodies, which later on will be recycled, and turned back to matter # 1.

The original size of a BS from the Big Explosion does not change to its "death," once they become matter # 1. Starting from the smallest black sphere to the size of the Earth's of 1,220 km diameter, the accumulation of "matter # 2" is directly proportional to the size of the BS, but when a BS is bigger than that the "game" changes. With Venus the rule becomes different: The bigger the size a black sphere has the less matter number two is tolerated, and eventually accumulated on top of it. These increasingly larger black spheres create the so-called "brown dwarfs" until they are so big that no material of any kind is tolerated on top of them, and they become stars. Venus is the smallest brown dwarf in the universe. Then comes Io, which outer size is smaller than Venus' outer size, but do not let that mislead you. Right at the border where the sizes of the BS are bigger than 1,220 km diameter and all such celestial bodies qualify as brawn dwarfs, some peculiarities arise. First is the outer size of these celestial bodies: Venus is bigger than Io, but Io is stronger. Second, Jupiter has this enormous magnetic field; which is the reason for enormous atmosphere, but is smaller in size and weaker than Saturn which almost lacks magnetic field but is stronger.

Because Venus is a bit smaller in size than the Earth, the scientists are assuming that it also has smaller mass, and that where they are wrong as well. Venus definitely has bigger density than the Earth, and therefore bigger mass. Venus has way bigger volcanic activity, and that could mean only one thing - it has a bigger black sphere inside than the Earth's. So the size of a planet or a moon is definitely deceptive factor of how dense and strong these planets, moons, and brown dwarfs could be.

Why our black sphere is the right size for the Earth? If we hypothetically switch the black spheres with Venus, life would not have developed ever on the Earth. The Venus BS is too big and therefore too powerful. The enormous output of volcanic gases would have overcome our atmosphere and make it too toxic for development of life, and these constant eruptions would not have allowed water on its surface because of the 80 C heat.

There are several questions that are puzzling to me, and need answers and explanations. First one is: Does a "planetary" working BS accumulates "matter # 2" releases BBPA to the very end of its existence? New material should be created and the overall size of this celestial body should grow to the very end of its BS activity. Such should be the case for the Earth as well. Through the volcanoes the new material adds to the crust and increases the size of the Earth. If this is the case then the Earth has been growing in size very slowly throughout its existence, and is growing as we speak. That is the reason why historical sites have to be "dug-up." The Moon had been growing in size until the "death" of its BS. Its lava had been flowing more toward the Earth, and that is why it has this bulge.

Venus has bigger and more active black sphere than the Earth, but is smaller in size. The reason for this is that Venus' higher temperature "evaporates" more material; therefore, its rocks should have different composition than the Earth's. The bigger size of the black sphere inside Venus indicates the first signs of slow "rejection" of the accumulated substance made of atoms and molecules on surface of its BS, and this might be interpreted as an "initial stage" of creation of a brown dwarf (BD) – call it a hybrid between planet and BD.

The answer to when our Moon's BS died should be found in the answer of when the volcanic activity ended? Somehow I know when Mars' BS died, thanks to scientist's report of the age of its biggest volcano. It died 30 million ears ago under the biggest volcano in the Solar System Olympus Mons. Another unresolved question is how a "dead" BS that created a planet or a moon looks like? We have the answer here on Earth - I have seen on TV stone balls made probably by the smallest BSs.

Hawkings reached the conclusion that through radiation the BS melt like ice. Of course his conclusions are baloney, for a simple reason, the guy is too "smart" with a minus sign in front of it, but I have to address it, because he lied to the world long enough. The question though remains: As the black spheres release their basic building particles of the atom with which they are packed, most definitely they lose mass, but are they loosing volume as well? If a BS melts like ice, it will get smaller and smaller, and there are reports that the Sun is loosing volume, then should we assume that 4.6 billion years ago it was way bigger than now? I doubt that. The size of the Sun is like a "box" to the end of its life. Allow me some speculation; it might be like a heart, pumping out the BBPA, but in a long run remains the same size. Let's say that the Sun is like a box outside, and we are looking at this imaginable box that 13.8 billion years ago was "full". After pumping out BBPA for all this time how much is left? The scientists have filled it with hydrogen gas in their imagination, and are telling you

86

that it might last another 5 -6 billion years. Would you believe them? ~~things are hopeless. Think about the Sun, it has been full of BBPA, but has been pushing them out in the last 13.8 billion years. After all no matter that the amount of BBPA was initially enormous, it is finite and one day will get to zero and stop working. So if the Sun is like a "box," and I believe it is, then we have no way of knowing how much "fuel" is in it, we might have hard time determining when it would be completely empty without knowing the "symptoms" for that.~~

~~Today's scientists are giving the Sun 6 billions years more of life. Sure enough they do not know what they are talking about. Their conclusions are based on two erroneous assumptions. The first one is that the source of energy for the Sun and any other star is a nuclear reaction where hydrogen creates helium in the center of it. This nonsense was invented by Eddington, and for some mysterious to me reason accepted by the "gullible" scientific community as "truth". Let me try to set the record straight: "Dear scientists, you've been lied to." The second misunderstanding is that the~~ The scientists ~~y~~ do not know about the UNIVERSAL CYCLES, described for the first time in this book. The "local cycle" dictates the fate of our cluster of galaxies. These "cycles" are uniformly the same, but have different time tables for different clusters. Our local group of galaxies currently is "scheduled" for the local "crunch."

Let us address the reported Venus' retrograde rotation. By now, you know that "red sift" is caused only by dust. Let us read this quote and you might stop believing that Venus has retrograde rotation around its axis. "Radiation – including light – reflected from Venus's approaching hemisphere is blue shifted, whereas radiation reflected from the planet's receding hemisphere is red shifted. By measuring the wavelength shift across the planet, astronomers determined that Venus's rotation is retrograde." This quote is from textbook for students 'Universe" Kaufmann 1994. This means that the reported "retrograde" rotation of this planet is based on Doppler's BALONY! This is something elementary – a rotation of a planet around its axis. Probably the same faulty method has been used to determine the Mercury's "retrograde" rotation? ~~These 600 pages textbooks for brainwashing young people are 90% nonsense.~~ This quote is from a 600-pages textbook for universities. All textbooks have to be thoroughly examined because 85% of their content is ~~My book has to be at least 1,500 pages or more to address all inaccuracies,~~ fake formulas, adjustable tailor-made fake mathematics, and piles of wrong "suggestions" – taken as "scientific truth" in the recent centuries.

THE EARTH

Yes, the truth has to come out, and I have to report it to you, but what if the news is not good. You know about these fishing boats with enormous nets that surround the fish. Let us say that one fish which represents the human race is in the middle, and have no idea that enormous net is closing around it, and there is no escape. We ~~are~~ might be in such precarious situation right now. The reversal of the magnetic poles does not pose danger to our lives. When imagine this process let us keep in mind the difference between axis of rotation and the axis of the magnetic poles.

What poses the real danger is the displacement of our BS from the center of the planet. Things are very simple; this displacement already has "damaged" our magnetic field. This "nature-made" "dynamo" works properly only if the BS has to ~~which is the core~~ stays in the center of the planet, and at this time it is not. The displacement of th~~e BS~~is ~~"core"~~ 1,220 km is 400 to 500 km off of the center in direction of Indonesia. The logical questions are: Why this is happening? What causes this? How long it would stay in this "off" position? Would it get any worse? Would it return back to normal? Keep in mind that the leaking of our atmosphere is caused by this, and the rate of air loss is directly proportional to the distance of displacement. All of our troubles of forest fires and global warming are caused by this displacement. Do the leading scientists know about it? Let me guess; THEY DO NOT HAVE A CLUE!

~~Because of lack of pertinent data, I have to speculate. The answers of how bad all these new phenomena would get relies in how close this thing would get to us; the closer – the worse to catastrophic. What could be the value of knowing how close and how long it would stay? What if we predict crop failure for 30 years? Then we can save food for this period, but if we do not, as is the case right now, we might perish out of stupidity.~~ To meet the challenges of tomorrow, we have to know what is in store for us. We have to monitor closely the movements of our BS in order to determine wh~~o~~at causes it, and from where this bad influence is coming. ~~As a scientific investigator on this matter I need it monitored minute by minute the movements of both magnetic poles.~~ This will point in the direction the disturbances our BS gets, and hopefully would lead to the discovery of the intruder.

Back to the Earth, which is the biggest "livable" planet possible in the Universe, and if anyone tells you anything different – he does not know what he is talking about. What dictates the size of a planet or a moon is the SIZE OF ITS BLACK SPHERE INSIDE. The Earth BS is the biggest possible for a living planet, because if it w~~ere any~~as bigger it would have produced brown dwarf, and that process have began with Venus. About the Earth is written quite a lot. The only news on my part is the presence of live black sphere in~~side the middle of~~side it. The fact that our BS is alive at this moment is the best news for our existence. Thanks to it we have living atmosphere and plenty of water~~;~~, and that guarantees our survival. The Sun and Earth's BS ~~core~~ are made from the same substance – BLACK SPHERE SUBSTANCE which consists of BASIC BUILDING PARTICLES OF THE ATOM. As the Sun has its "Solar wind" of particles (BBPA), the Earth's BS has the same "wind" of BBPA but in a very ~~low~~small scale, and that is good news for us, because our way-smaller "BS" TOLERATES THIS METAL AND ROCKY MATERIAL ON ITS SURFACE - SO WE CAN LIVE ON IT! The moderate temperature coming from inside allows the creation of crust. Without this life-saving crust no water, plants, and animals are possible. The moderate activities of our BS also contribute to "tolerable" amount of toxic gases coming from the universal volcanoes. The BBPA released from the Sun go in free association and creation of elements and chemical compounds in open space, but the same process is going on inside any living planet or moon, and chemical elements and compounds are created in the mantle, but some of them are pushed to the surface as gas, because of the high temperature, and that are the VOLCANIC GASES. This is valid only for the "universal" and not for the "subduction" volcanoes, which gases are from coal, oil, natural gas, etc. Between planets and moons there are no substantial structural differences, but between planets and brown dwarfs there are.

    The amount of gas that is~~has to~~ com~~ing~~e from the Earth's universal volcanoes is not too much, and overall our nature and atmosphere manage to neutralize it. The surface of the planet has the accommodating temperature range in order for surface water to be present. Underground water is possible because the heat under the crust is not that high, if we compare it to Venus' 480 degrees C. The source of water on the Earth is of a local creation. Definitely through BS activities water is created.

    The increased carbon dioxide coming from human activities makes global worming worse, but is not the main cause of it., ~~but that is what the leading scientists are telling you, and I can assure you that they do not know what they are talking about. But help is in the way, I would teach them! All I want them to say is "uncle." And here what their response might be: "The uneducated idiot wants us to call him an UNCLE!?!" Jokes keep my sanity.~~

—Supposedly we are in this "switching" period, and~~but~~ such "period" has not happened in the recorded history. The atmosphere is already leaking into space, and the magnetic field is weakening.~~.~~ Scientists are assuring us that the amount of air we are loosing is small, but let me give you a secret – they do not know what causes it, how long would last, and would it get worse? The condition of the magnetic field is directly linked to the displacement of the BS. No one can predict how bad the weather would get, because they do not understand the weather processes as well. Due to this leaking, the parameters that determine the weather patterns are changing. How bad it would get depends on the "severity" and duration of this leaking. I know what they are doing – looking at their super dumb computers, and receiving some half-cooked answers – "nonsense in; nonsense out." The computer is as dumb as its programmed data, and we know that its content about Astronomy is 90~~85~~% pure "scientific" TRASH!

    When it is at its rightful place which is the center of the Earth the BS is doing its job of generating quality magnetic field, and the farther it "goes" from there - the more "chaotic" and less working this field gets. There is one of my rules: "No magnetic field – no living atmosphere," and there is another rule: "The farther the BS gets from the very center of the Earth, in worse condition the magnetic field would get; which would lead to bigger volume of atmosphere per unit of time loss;

~~which would lead to~~ farther shrinking of the atmosphere; ~~which would lead to~~ more severe bad weather; more forest fires and bad hurricanes." And there is another very disturbing reality: Because the BS is 700 denser than the surrounding rocky material, the Earth would wobble increasingly worse when getting farther from the center, and might reach the point of "no-return," which I do not know where ~~is~~ it might be at this moment, but Mars' BS could not return back and for the last 670,000,000 years of its life remain close to the surface of the planet where it "died," but if there were any Martians they died 700,000,000 years ago. Explanation of "nuclear reaction" elements reported on the surface of the planet. When a BS releases BBPA a SPECIAL NUCLEAR REACTION occurs. The same thing is going on inside the Earth, but is some 3,500 km under our feet. ~~Do not look for their bones in "Tharsis Montes."~~

If you know the scientific truth, you might be able to protect yourself, but with their scientific nonsense, may I suggest, that you might be doomed. Advice to countries like Nepal, Tibet, and cities at high elevations like Quito, Bolivia – get to lower grounds, you might not be able to breath. ~~have hard time breathing.~~ Advice to the climbers of high peaks: The air has gotten thinner, and this process is on right now. Same advice goes to the pilots in these altitudes.

~~Black spheres have their "communications." The scientific community does know about its existence because do not know and understand black spheres. First they do not know that BSs are inside planets; second do not know what their density is; third do not know their "communications and laws of command and control." Consequently, they are not considering the development of devices that can "read" these "communications."~~

OUR MOON

Our Moon's BS is dead – when that happened and why so early is unknown? The BS of any celestial body "dies" with the last volcano. How black spheres inside planets or moons die, and what happens to their "bodies" after "death?" The Earth's BS is 1,220 km diameter, if it dies what substance would remain in its place? This question is easily answered because there are small spherical objects here on Earth, which have been created by BS inside. The "death" of a BS is caused by material exhaustion – all the basic building particles of the atom had been pushed out, and that spells end of operations, but there cannot be some empty spherical space left in the center of the Moon. Most likely the "dead" BS becomes "matter number two" – atoms and molecules, but what chemical elements? My guess is rocks and iron, but farther investigation is needed.

Let us look at what nonsense the scientists believe. According to Smithsonian Universe 2020 "the Moon is too small to retain substantial atmosphere." The Moon cannot have "substantial" atmosphere because of its size, but it is not the size and the gravity that holds an atmosphere but the magnetic field which the Moon cannot have naturally because the Earth or the Sun "prohibits" it from rotation – this is the law of the "systems" "planet/moons" or "brown dwarfs/ moons. This is a universal law and is very simple: In every planet/ moons or brown dwarf/moons system; the moons are not allowed to rotate around their axes, therefore to talk about "living atmosphere" is unthinkable. ~~And that shows the total ignorance of the leading scientists.~~ The existence of living atmosphere on the Earth depends 100% on the well-functioning magnetic field~~, and this is another ignorance on their part, which is criminal in today's circumstances~~. Live BS and rotation are essential requirements among others to have a living atmosphere. We can make the conclusion that no moon in the universe could have a natural living atmosphere – therefore no surface water. We know that Saturn's moon Enceladus has surface water which is protected by ice. Water is easily self-created and is abundant throughout the universe, but unprotected by a living atmosphere it is easily blown away. In Enceladus' case water is protected by ice.

The current terminology of so-called "synchronous" rotation where only one side of the Moon faces the Earth is baloney – the real terminology should be: "Rotation is prohibited!" It has been already established that the rotation of the planets is commanded by the Sun, so the prohibited rotation for the moons could be combination from the planet, ~~most likely comes from~~ the Sun, and the system ~~as well~~. It could be a "system" requirement. ~~The whole thing is one giant system where smaller systems are inside larger.~~

The scientists are looking for potential exoplanets that might foster life and water. There cannot be surface water without living atmosphere, and we know that the guarantor of this is living atmosphere and ~~is a~~ magnetic field. The other requirements are right size of BS and the right distance from the star. What if the BS is not at the center? There cannot be a bigger PLANET in the entire universe than the Earth. (It could be "bigger," but very insignificantly.) Anything bigger is a BROWN DWA~~A~~RF, and the funny thing is that some smaller than the Earth spherical bodies are brown dwarfs as well – Venus and Io. What maters are the size of the BS inside, and not the visible outer size. Too big and powerful BS will create too much pollution for living atmosphere, and too high surface temperature for water to exist. Scientists are talking about potential planet with size 10 times the size of the Earth – no such "planet" can exists ever– that would be gassy brown dwarf, which basically is a "living hell".

Although the black spheres of the Earth, the Moon, and Mars are from the same black sphere substance, and began their functioning simultaneously after the Big Explosion 13.8 billion years ago, some are "dead" and others "alive." Why the Moon's and Mars' black spheres died early, but the Earth's is still functioning? Obviously the black spheres of planets, moons, and small stars "die" when the big stars never die. Here is a question concerning our very existence: How long our beloved BS inside the Earth would function? The answer is simple: It would function until all the BBPA are discharged, but how we could know what supply of

BBPA is left? Probably there is a limit how low on density a BS could go before it dies. So our BS has an approximate density of 2,365 tons per cubic meter, as this number would go down, it would eventually reach the number when the BS could not function any longer, but we do not know this number yet.

Smithsonian 2020 page 137: "No one knows exactly how the Moon was formed." I know, and I am about to tell you; VERSION NUMBER ONE: When the formation of the Solar System began 13.8 billion years ago, and according to how the black spheres work, the Sun ordered the biggest BS in this orbital spot to orbit around it, and this was the THIRD available orbital distance from it. Happens that the Earth's and the Moon's BS were there. At this time the BSs of the Earth and the Moon did not have any matter # 2, and after their BSs become matter # 1 they were shining like little mini-stars. Their sizes were 1,220 km and 340 km diameter; these were only the BSs of the present day Earth and the Moon. Through the activity of these BSs matter # 2 was created and accumulated on their surfaces. The Moon's BS did its job get all the BBPA out and "died." The Earth's BS is still creating new matter # 2, and in the process increasing the size of the Earth and decreasing its density.

VERSION NUMBER TWO: This version is more likely. As we know, stars get compacted by supernova explosions which are not as powerful as the Big Universal and Local Explosions that create galaxies. The conclusion is inevitable; initially all potential BSs in these galaxies were neutron stars and white dwarfs consisting of matter # 3 which becomes matter # 1. To me this process of matter # 3 becoming matter # 1 at this time is not quite clear. We see how white dwarfs "steal" matter # 2 from other stars in binary systems. That tells me that they are "thirsty" for some particles which they lost in the explosion. BBPA are indestructible; and "live" in this "perpetual motion; "WHATEVER EXPLOSIONS MIGHT HAPPPEN; THE BBPA ALWAYS MANAGE TO LAND ON THEIR FEET!

Let us look at the creation of the Earth and Moon under the second version. For sure the fore-mention sizes of their BSs were 1,000 times smaller. Slowly and how exactly unknown they gained their real sizes; thus becoming real functioning BSs, and the creation and accumulation of matter # 2 began.
— The Sun in this instance acts exactly like the Central Black Sphere that commands the Galaxy. The orders go to the biggest black sphere. The Earth took the order and occupied this predetermine third orbital distance. The Moon happens to be in the Earth's field of influence and accordance to the universal laws of subordination as a smaller BS began to orbit the Earth. Let us not forget that some

ratio requirements had to be met; the Earth/Moon BS ratio is 4 to 1. ~~Keep in mind that the size of the Moon looks quite big for size of the Earth, so obviously we have some limit as to how big of a moon certain celestial body can handle. In the case of the Earth/Moon relationship we have 1,220 to 340 which is around 4 to 1.~~ Here is a rhetorical question: Could the Earth have become a moon? The answer is yes. The Earth could have become moon to any of these powerful "gas giants," but could our planet have become moon to today's moon Io? If the Earth and Io were back then next to each other, as the evidence show the Io's BS bigger than the Earth's; then Io would have taken this orbital spot. That does not mean that the Earth would have become Io's moon, because Io might be too weak to handle the biggest planet in the universe. Mars' BS size was 678 km. Mars could not handle our Moon, but it could have bigger moon than these two rocks going around it, and we if apply the ratio of 4 to 1 ~~would works~~; then Mars could handle moon as big as 1,700 km diameter. We have an interesting phenomenon here. Initially the Mars' BS made these stones orbiting it, but now when the BS no longer exists they continue to orbit the planet, why? The power of the BBPA "influence" has been transferred to the "matter # 2." The compact mass of the black sphere created the planet Mars (around 10 times larger), and here we have the case of "transformation and transfiguration." The basic building particles have transformed themselves to atoms and molecules, and the black sphere has transfigured itself to ~~10 time~~ 10-time larger celestial body which is the current Mars. In this case we could safely assume that this "influence" is property of the particles and not to the atom. This "influence" of these particles whether they are in BS or in "matter # 2" (atoms and molecules) is forever with them. Matter, energy, and influence are indestructible and inseparable aspects of the BBPA. Maybe we could say that the ever-present "influence of the BBPA is a manifestation that matter and energy are inseparable.

PHONY COMPUTER GENERATED TELEVISION GARBAGE

I am leaving this as a home-work for the reader. On the TV screen they are showing you computer-generated nonsense, portraying it as some sort of "reality," where some stones looking like red hot charcoal are hitting each other and creating some mythical "planeticimals." Because we know that the scientists are 85% brainwashed, we could say that all computer generated images and calculations are fake.

MARS

700,000,000 years ago Mars' BS was dislodged from the center of the planet. It never returned back to where it belonged. It went all the way to the surface of this planet, and stay there for 670,000,000 million years; through its activity it built 4 of the biggest volcanoes in the SS, and 30,000,000 ago it died. Why these volcanoes become the biggest in our SS? That is because the BS was right under them pumping all the gases and material into one volcano at the time, and a. At this time there was no longer atmosphere to wear them down. Mars did have living atmosphere and plenty of surface water, but 700 million years ago after its BS was forcefully removed from the center of the planet by an intruder like the one that threatens us now; it lost it all; first its magnetic field, and then its atmosphere. According to one of my Popov's universal laws; "if there is no magnetic field; no formation of living atmosphere could be is possible." What caused the removal of the BS was a strong and contradictory dictory to the Sun's "influence" from another powerful celestial body, which was powerful enough for the weaker Mars. According to another of my Popov's universal laws; "In the Universe, the influence of a celestial body increases with the amount of BBPA inside it." The most abundant is matter # 1 which is in all stars. The rule, the bigger the size; the bigger the influence applies for them, but with "white dwarfs" things change dramatically. White dwarfs are 1,000 more powerful than stars. They are severely compacted, and do not act as normal BSs. They are made of BBPA, but are not arranged in the "special way" like in the BSs; therefore they do not act like them. According to another of Popov's universal laws; "The smallest subatomic particles are indestructible, and inseparable from their energy and influence." That means that regardless that white dwarfs are "victims" of explosion; the power of their influence not only remains with them, but it is 1,000 times multiplied. 700 million years ago Mars suffer "mortally" from some unwelcome influence, but thankfully did not affect the Mother Earth! Mars had a working BS as recently as 30 million years ago, but it lost its atmosphere 700 million years ago. For 3.9 billion years prior to that, Mars had surface water and living atmosphere. Let's remind ourselves that without living atmosphere there cannot be water on the surface of any planet, because liquid water on the surface requires narrow range of temperature which only "sheltering" "livable" atmosphere can provide. Obviously ice can create protection for existence of water, but this water cannot be clean as the surface water on the Earth. Water is easily created by all BSs, and that is why it is quite ubiquitous all over the universe. What the scientists should look for in quest for liquid water on the surface of exoplanets? First of all, potential "livable" planet cannot be bigger than the Earth, and these small sizes they cannot see them any way. What they see are brown dwarfs, because they have bigger sizes, but there cannot be life on them because of high temperature. Another thing they have to look is well-working magnetic field, because it is the guarantor for living atmosphere, which on its part is guarantor of liquid water on the surface. Then look at the size of the planet — should be smaller or equal to the Earth, but Io is smaller also, but is a brown dwarf! So the question is how to determine the size of the BS inside. Scientist are talking about planets 10 times bigger than the Earth — no such "planet" can exists, if a celestial body is that big, it is a brown dwarf, or a "planet" that could exist only in their uninformed heads. Even if Venus were to be

~~where the Earth is, due to its "too big and overactive" BS producing the high surface temperature and toxic gases — no living atmosphere and surface water would have been possible ever, no matter what the distance to the Sun is.~~

Today scientists do not know what causes these reversals of magnetic poles, and we happen to be in the "middle" of it. Another thing that scientists do not have is a "tool" to get into black spheres communications, namely CBS to Sun; Sun to planets: and planets to moons, and most importantly "intruders" to "established solar systems." These communications are part of the "Black Sphere Influence and Communicational Orders," but the scientists are stuck with Newton's nonsensical and no existing "gravity." If this tool was available right now, we would have been able to detect the intruder.

Mars' BS ended its life under the Olympus Mons 30 million years ago, but I was intrigued by a report that on Mars was found chemical element that could be produced only through nuclear explosion. The creation of "matter # 2" is a "special nuclear reaction" which ~~probably had~~ produce~~d~~s these element~~s~~ which went to the surface of the planet, because the working BS was right under ~~it~~ the surface of the planet. Why we do not have these elements on top of the Earth is, because the~~i~~se special nuclear reaction~~s~~ occur~~s~~ about 3,54,000 km under the surface of the Earth. The "death" of Mars' BS came under the biggest volcano in the SS Olympus Mons. ~~In order to confirm this theory, the biggest concentration of this element should be in the vicinity of this volcano. I do not know what this element is, but if it could be dated back to more than 30 million years ago, then this theory might be true.~~

The scientists believe that the stars are going through "evolution" since the NONSENSE called Big Bang, and yes the stars are going through evolution, but not through their version of events. All that the stars are doing is releasing BBPA, and in reality "matter" lives forever, but in their heads matter turns to energy, then this energy is forever lost, and one day the sky would be dark. ~~So they live in some world of their "invented" nonsense upon nonsense, which has nothing to do with reality.~~ The human race admires this fake formula $e = mc^2$, and makes the argument that mass of two hydrogen atoms ~~have~~has more mass than the resulting one atom of helium - therefore there is missing mass that supposedly has turned to energy. The~~By the way our~~ scientists "see" that after all these supernova explosions the stars remain~~s~~ intact, but they still call them "dying" stars. So they truly believe that these stars have run out of energy forever. All these deceptions come from series of wrong assumptions piled-up in the last several centuries. They disregard the evidence in favor of their fake theories. Join me to say NO to the PILE OF NONSENSE, and YES to the TRUE SCIENCE WHICH SHOULD BE OBSEVATIONAL AND BASED ON FACTS – not on "suggestions." ~~(Eddington only SUGGESTED that stars are initially filled with hydrogen gas), and crooks and liars like mentally retarded Einstein and Hawking?~~

ASTEROID BELT

From Smithsonian Universe 2020 page 170: "Asteroids are remnants of a failed attempt to form a rocky planet that would have been about four times as massive as Earth." This is a ridiculous statement. They even know the size of the planet! I have to go and find out how they came to this FAKE conclusion, but there is danger of going through so much of their "scientific garbage" one might die from toxic fumes ~~of their conclusions. There cannot be planet bigger than Earth!!!~~

There probably was a planet there, because this "asteroid belt" is preordained "orbital distance" for a planet, but this planet was destroyed or flunked out by Jupiter, probably with its moons as well. Jupiter is just too powerful to occupy only one allotted orbital space where the spaces are "narrower" on the side of the Sun. The orbital spaces are increasing larger from the Sun to the periphery, and all the big and powerful brown dwarfs are in the last 4 spaces, but obviously the distance between fifth and sixth space is not large enough to accommodate the big and powerful Jupiter. I was wondering how Saturn does not get affected negatively, but when I read the information about the size of Saturn's "solid core" ~~BS~~ being bigger than Jupiter's, that told me something else: Saturn is a bigger bully than Jupiter. Other interesting information is that one of the Saturn's moons is orbiting at 25,000,000 km from it. Compare that with Earth's distance to the Sun 150,000,000 km. To understand how powerful Saturn is, imagine our Moon going around the Earth at this distance.

Here is statement about the structure of the asteroids from The Universe by Smithsonian 2020 page170: "The radioactive decay of elements within the asteroidal rock melted these large bodies and, during their fluid stage, gravity pulled them into spherical shape before they cool." Yes, good explanation for 6 year-olds. Notice, they need the so-called 'radioactive decay" to melt the interiors or planets, moons, and asteroids, but cannot account why some planets and moons are "dead" already and some are "alive." So here is their profile of what arsenal of knowledge they have, and how they make their phony conclusions. If something inside a planet or a moon is melted therefore is done by radioactive decay, or this heat remained there from the "mythical" Big Bang 13.8 billions years ago. So here is my joke: "Please, make me one real hot coffee." "How hot do you want it?" "So hot, that should remain hot in the next 13.8 billion years."

Here are some of their explanations for volcanoes. One textbook was explaining existence of volcanoes this way: imagine that you are walking on mud bear-footed and the mud is squirting between your fingers. Next nonsensical explanation is called "ice volcanoes" on Saturn's moon Enceladus. The reality is that water spray from this frozen moon is due to volcanic eruptions under the thick layer of ice which was melted by the "universal" volcano under it. Welcome to their "gravitational" volcanoes occurring right now on Jupiter's moon Io happening because of two opposing "gravitational pulls": On one side Jupiter's gravity and on the other side the moon Europa's gravity – COMPLETE NONSENSE! In this book you are learning the "scientific truth" about volcanoes. Newton's no existing gravitation is doing outstanding job of many nonsensical explanations, and even it holds the entire universe together.

By now you KNOW why there are spherical asteroids, and why all celestial bodies are spherical. All black spheres through releasing BBPA are creating proton, neutrons, atoms, and molecules. This process creates NUCLEAR energy which is the heat which meltsing the interior (the mantle) and because of gravity the outer shape is spherical like the BS under it. For the first time you are hearing a coherent explanation of why planets and moons are spherical. And if Michio Kako claims that his explanation was the same, let me remind you that he does not have the foggiest idea about BSs present inside spherical bodies; their tolerance to atoms and molecules on their surface; why the iron is first on the surface; what processes pile this matter on top of them; how the magnetic field is generated?

Undeniably there were more collisions in the Solar System before. Whole planets and moons can be totally destroyed if the impact is strong enough. This asteroid belt is very revealing and supportive for my theory of "commanding" influence rather some gravity. The orbit of this belt is predicted by Bode's law which confirms my theory of Sun's "predetermined" orbital distances for all planets. These predetermined orbital spaces are like TUBING. The hold of the Sun of a planet being there is so STRONG, that if external influence tries to pull this planet out of this tubing, the planet would be destroyed in the process. Why the BS is treated differently than the mantle; remain to be solved.

For the Earth's orbit it is $(6 + 4)/10 = 1$; for Mars' orbit $(12 + 4)/10 = 1.6$; for planet X which is missing and where the belt is located $(24 + 4)/10 = 2.8$. Kepler third law which states that the "average orbital distance" cubed should be equal to the time of "one complete orbit" squared not only predicts the distance from the Sun of this belt, but the speed of this debris: 2.8 cubed equals 22.4, therefore what number squared equals 22.4 should be around 4.8. Reportedly the asteroids are orbiting the Sun between 4 and 5 years. So the Sun has captured and "ordered" these stones to be in this prescribed orbital distance – destinedthat is for a planet and "ordered" them to complete one orbit between 4 and 5 years, but there is more – the Sun MADE THEM ROTATE AROUND THEIR AXES!!! Through this belt I made the discovery about who makes the planets rotate around their axes – the Sun does. This discovery answers the question: What makes Mars rotate with such precision, even after its BS is dead, so it does not matter does some planet has live BS inside or not, if the Sun is functioning all planets would rotate!

So the electricity necessary for magnetic field is not generated through rotation of the core the way we make electric motors, but through rotating the "casing" which is the metal going around the BS. In this case the BS should be "prohibited" from rotating, and stay -stationary. with the rest of the planetary mass, and in

~~order to account why the metal is first next to the BS; my contention is that except from the regular gravity, iron is subjected to additional "gravity" which is magnetic by nature.~~

~~Scientists are trying to use Kepler's third law on the Galaxy, and this is a mistake. If it works for the Solar System, it does not mean that would work for the Galaxy. The central black sphere (CBS) in the center of our Galaxy "commands" differently all stars, planets, and dust present inside the Galaxy than the Sun. These two black spheres have different tasks; they have different sizes, obligations, and different approaches. In Kepler's third law the distances of all planets are "tied" to the Earth's distance to be ONE (1.0). HOW IN THE WORLD, THE SCIENTISTS ARE DETRMINING WHERE IS THIS DISTANCE OF "ONE" IN OUR GALAXY? The distance of "one" in the Solar System happens to be the Earth's distance, and remember, after it is cubed should be equal to one, and the time going around the Sun also should equal "one" when it is squared. No one understands that now and in the last 300 years starting with the "great mathematician" Isaac Newton.~~

GAS GIANTS

The 4 gas giants in my classification are definitely brown dwarfs. As such they are way more powerful than the 4 rocky planets. It is certain that all rocky planets can be moons to any of these giants. The giants needed more space to exist, for example the most powerful of them is Neptune, and has a moon with orbiting distance at 48,000,000 km, and occupies the last orbiting space which is the widest, and is almost one half of the entire planetary field. Jupiter is the weakest, and occupies the one in the middle between the small rocky planets and the 4 giants. Here is the question: Did this arrangement happen by chance or the Sun and the "system" had something to do with it? My answer is that the Sun "made" this arrangement. It cannot be by chance that the smallest planet Mercury should be next to it and the most powerful should be at the periphery. Or should we believe that the planets jostle for orbital spaces, and the current arrangement just happened by chance? Should we reach this conclusion? ~~Here is unproven law of the Sun's "duties" and "ability:"~~ "In every planetary system the central star 'pushes' the strongest celestial bodies to the periphery, and keeps the weakest to itself?"~~.~~

~~SMALLER BROWN DWARF'S POWER AND SIZE~~

~~Here is Jupiter, the weakest of the 4 gas giants, but with the biggest outside appearance! The power of one celestial body is determined by the size of its BS. For the rocky planets and moons at this stage of evolution from the creation of our Galaxy, when you look at their sizes, you can know the sizes of their BSs which are 10 times smaller. This is the case with the Earth, but to determine the size of the~~

JUPITER

The scientists are who do not know what the nature of all celestial bodies is, and with my new discoveries it is only "natural" that new and updated definitions have to be made. By the way they classify "brown dwarfs" as "small" "failed stars." There is nothing failing in the BDs simply they are smaller than the smallest stars. The difference between stars and brown dwarfs is that the last begin to accumulate matter on top of their BSs. The first matter accumulated by the biggest is gasses, and the smaller sizes accumulate more and more matter # 2 until they qualify as planets.

The classification "brown dwarf" for celestial bodies is between planets and stars. That means the larger once are very powerful. They are as powerful as stars' pound per pound, because some of them do not have any metal or rocks on top of them. But with the smallest sizes we have an interesting phenomena – initially their outer sizes are "getting" smaller than planets. Venus and Io are smaller than the Earth but more powerful, yet the Earth is the biggest "planet" in the universe.

Smithsonian 2020 reports temperature of 5,000 degrees C close to the core for Jupiter and 4,700 C for the Earth. I disagree with these numbers. Jupiter is way more powerful and should "generate" higher temperatures that the Earth. These temperatures are highly inaccurate in numbers and "location." There is a simple reason for that; they do not know at all how and where this temperature is generated. The temperature comes from THE SPECIAL NUCLEAR REACTION when BBPA create "matter # 2." Because Jupiter's BS is too powerful, it evaporates more rocky material, but we should note its very strong magnetic field, which might be as a result of big amount of iron?

If we look for guidance where ~~it~~ the highest temperature occurs~~,~~ when looking at the Sun; it is in the corona ~~, the highest temperature of~~ (2 million degrees C) ~~happens in the corona~~. Proportionally inside every planet and small BD the highest temperature occurs in similar distance from the inner core, where IRON IS ~~all of "these"~~ CREATED ~~IRON~~ which returns back on the surface of the BS where the temperatures is lower and stays there in the case of the Earth, but cannot stay there in the case of Saturn, where ~~Probably~~ the higher rate of BBPA release "expels" the iron. And let me repeat again that the temperature of the very surface of any BS including the Sun is insignificant. That leads me to the conclusion that ~~no~~ "heat" is NOT ~~is~~ coming out of the Sun. So here is the question: If a nuclear reaction creates heat at the center as they claim, then what is this "puny" temperature at the Photosphere of 5,700 C and 3,000 C at the surface? ~~,~~ ~~a~~And why "mysteriously" would get to 2 million C at the Corona? These temperatures support my theory 100%!

Saturn, Uranus, and Neptune have a different regulating mechanism than Jupiter. They should have some equivalent of "solar wind" in a smaller scale proportional to their size. My bet is that the scientists and the observers of the 4 ~~se 3~~ gas giants were not looking for this "wind of particles" because they are not aware of it.

Hydrogen and helium are the two gases most widely available in the universe because are the easiest to self-produce, and naturally the two gases are predominant in Jupiter's and the other gas-giants atmospheres. Here is "Smithsonian's" 2020 guess on the composition of the Jupiter's core – ~~e:~~ "solid core of rock, metal, and hydrogen compounds." FOR THEM EVERYTHING CONSISTS OF MATTER # 2. Also, they think that at the core of these giants there is ICE, in the same time they see and report that there is internal source of heat. They do not know where this heat is coming from.

When the scientists are talking about a possible "planet" 10 times bigger than the Earth, they are talking about brown dwarf, because such huge "planet" does not exists. They do not have a definition for a "living planet," but I have: Living planet not only has to be in the right distance to the star, but it has to have the right size BS which diameter should not be bigger than 1,220 km, and it should "function" properly in the center of the planet in order to generate magnetic field. A "quiet neighborhood" around this "solar system" is also necessary. We have an "intruder" in our Solar System, and with its "influence" which is contrary to our Sun's "influence" threaten very seriously the "livability" of our planet. I hope the Earth and we survive, but definitely it would not be easy. In order for us to survive we have to know what is in store for us. If it is up to me, I am pressing the "panic" button. My understanding on the matter is totally different from the "scientists," and we cannot be correct both ~~correct~~!

Here is a recap. Obviously some spaces "allotted" for planets by the Sun are taken by brown dwarfs, and all "gas giants" are brown dwarfs. Saturn definitely is stronger than Jupiter, but might not be the biggest BD in the Solar System, because Neptune has a moon with orbital distance of 48,000,000 km which is one third of Earth's distance to the Sun. The size of Neptune's and Uranus' BSs is unknown, and only when this is accurately reported, proper classification could be made. At this moment we can only speculate, and here it is: Jupiter is the weakest with the smallest BS from these 4 giants, and it is one giant balloon. Our observations are deceiving us – it has the biggest outer size and the biggest magnetic field, but these are signs of weakness. It has been "stated" that Saturn has 1.5 "solid core" bigger than Jupiter – that said it all – Saturn is stronger! I assume that Saturn's BS is bigger than Jupiter's.

Why Jupiter's magnetic field is 20,000 times stronger than the Earth's? Because a lot iron is produced on top of the hugest BS that still tolerates iron on top of it. Rocks evaporate faster than iron, plus the highest temperature on any BS is generated well above its surface in the mantle; therefore rocks are evaporated but the iron remains. Why Saturn has magnetic field only 0.7 from the Earth's, because the iron is almost missing. These differences are as a result of temperature differences of the two BDs. By the way, I believe that magnetic fields could be created only with metal when the scientists think differently. Keep in mind that they assume that at the core there is ICE!?!

What is interesting to me is that if any one of these 4 giants with their enormous power were in the Venus place, it probably would have devastated Mercury and the Earth. Imagine Neptune there with its RADIUS OF INFLUENCE of 48,000,000 km. Thanks to the remote orbital "spots" these giants have taken – there is enough space for them and the smaller planets are able to exist. Here are the preordained orbital distances for planets and brown dwarfs in our Solar System in millions of kilometers: Mercury – 58, Venus -108, Earth – 150, Mars – 228, Missing planet (asteroid belt) – 320, Jupiter – 778, Saturn – 1427, Uranus – 2871, Neptune – 4497. Notice how the small rocky planets have taken the "available orbital spots" where there is not too much space between them, and they can peacefully coexist, and the giant and powerful brown dwarfs are at these remote spots otherwise there would be no peaceful coexistence. One planet next to the big bully Jupiter paid the ultimate price – we do not know where it is or what happen to it.

Some of the gas-giants are mistaken as some sort of balls of ice and particles. The reality is that they are way more powerful than any rocky planet, but at the periphery of their atmospheres there might be ice. Remember, the celestial bodies in the universe are as powerful as its BS inside. Scientists are giving real low numbers of density for the gas-giants and the Sun. These numbers are absolutely "cooked" and have nothing to do with reality. How the scientists see the Sun? First they "filled" it with hydrogen, which is nonsense, and then they "calculated" its mass with some concocted and inaccurate formula from

~~Newton, who did not understand Kepler's third law. The mass of the Sun is calculated in kilograms and so is the cheese on Earth, so we can compare them, and the cheese weights more per cubic meter! Total and complete nonsense!!! I am repeating this, but I look at this fake formula where the distance from the Earth to the Sun is inappropriately given in meters, and the orbiting period in seconds, but there is more nonsense in this formula, and that is so-called "universal constant of gravitation." There is nothing "universal" in this lab concocted number derived from "matter number two," but is inapplicable to the 98% of the "matter number one" of the universe.~~

THE JUPITER'S MOON CALLED IO

The Jupiter's moon Io is the most interesting moon in the Solar System. Supposedly it is bigger in size than our Moon. Do not be fooled by the size. This so-called "moon" could have been one of the planets, and the most interesting thing is that it is actually a brown dwarf – more powerful than any of the 4 rocky planets. We can see its surface thanks to Jupiter's gravity that sucks out its constant emission of gases; otherwise its surface would have been covered with toxic gases like Venus. The biggest size for a black sphere inside a planet is the one inside the Earth, and Io's one is bigger – that qualifies it as a brown dwarf. If the Earth and Io were next to each other at the creation of the Solar System, Io would have taken Earth's orbital space, because this the "law" of the black sphere "influence." The "order" from the Sun has to go to the bigger of two, and in this scenario Io takes priority over the Earth. The question is would the Earth would have become Io's moon, and probably the answer is – no. The Earth might be too big for Io to handle it as a moon, but for sure any of the "gas giants" could ~~make~~handle the Earth ~~its~~like a moon. Remember, Io is stronger than the Earth and is a moon, so any celestial body with smaller BS inside than Io's can be a moon to Jupiter, and that includes all 4 rocky planets.

  Let's have some fun reading some scientific nonsense of today. Here is how the scientists are explaining the existence of volcanoes on Io. From Smithsonian Universe 2020 page 184: "Io orbits Jupiter quickly, every 42.5 hours or so. As it orbits, it is subjected to the strong gravitational pull of Jupiter on one side and the lesser pull of Europa on the other. Io surface flexes as a consequence varying strength and direction of the pull it experiences. The flexing is accompanied by friction, which produces the heat that keeps part of Io's interior molten. It is this material that erupts through the surface and constantly renews it." Now you know their explanations as to why there are volcanoes erupting on this moon. I am surprised that in this "scientific" explanation they forgot to mention what Pinocchio was doing – it is possible that he was helping create these volcanoes as well.

Let's recap this fairy tale. Once upon a time there was a guy called Isaac Newton. An apple had hit him on the head. He said to himself: "There is a force that pulls objects toward the ground," and his dog which was next to him also thought the same thing. At the same time monkeys in Africa "independently" of Newton also discover this gravity as well. Because we are smarter than monkeys and dogs, our representative Isaac Newton REASONED that if apples fall to the ground then the same force makes planets go around the Sun. At that time Samuel Becket was not around to tell him about one his characters which: "HE REASONED, AND HE REASONED WRONG," but even if this valuable advice had been given to Newton, the result would have been the same – he would have invented his gravitational nonsense. And that is how Newton discovered what the dogs and the monkeys already knew, as far as the Sun's gravity extending to the planets; he was 100% WRONG! At this very moment the gullible human race with all its scientists, professors, Ph. D. holders, super computers, and artificial intelligence believe that the universe is held together by "falling apples gravity." By the way, Newton's gravity is backed-up by mathematical chewing-gum gymnastics, which was brought to perfection by the biggest mathematical crook in the universe - Einstein. "Newton, thank you for inventing (sucking off your fingers) this wonderful and magical gravitational nonsense, otherwise what explanations we would have given to the children, and how we could be taking more money from others to explain to them what we do not know?". At this moment I have tears in my eyes. I spoke with the Little Riding Red Hood about this unfair "gravitational pull" on both sides of Io, and here what she had to say: "It is highly unfair how they are pulling on Io in different directions, and I totally support him in his volcanic eruptions." I wiped my tears. By the way, these are the same "scientists" that are informing you about the Sun, the Earth, the Earth's atmosphere, the changing weather, and the global warming, so go ahead and BELIEVE them. When you are listening to the news, and the anchor person tells you what the scientist's opinion on these subjects is – take it with a grain of salt.

SATURN

Solar System is not a planetary system, but a system available for planets, moons and BROWN DWARFS (BD). "Smithsonian Universe" 2020 is "informing" the readers that Saturn is so light that if we put it in ocean of water; it would float. Let me inform them: If Saturn is put in ocean of water; the ocean will disappear in a split second. Scientists are giving Saturn's mass to be 95 if Earth's mass is 1, but the Jupiter's mass they give 318. I would call this "a deception." Jupiter solid mass is real small 23,000 km diameter.

At what size of black spheres the brown dwarfs lose their magnetic fields? The answer is between 10 and 20 times the size of the Earth – Jupiter has magnetic field, but Saturn does not. All brown dwarfs smaller than Jupiter can have magnetic fields which power increases proportionally with the size of its BS inside, but there is a limit, once the size of the BS gets to be between Jupiter's and Saturn's – the magnetic field "disappears." Explanation for this phenomenon is the increased power of the BS and its intolerance to "matter number two" on its surface. The scientists are talking about "creation" of magnetic field in Jupiter by "metallic hydrogen," and I doubt that. Only iron can do the job.

NEPTUNE

Neptune is the most powerful "planet" (their definition) in the SS. It is a brown dwarf. Reportedly its "solid core" is 24,000 km, which is the same as the Jupiter's "solid core." From this arises the question, as to why Neptune should be way more powerful than Jupiter? The answer is very simple: Neptune has bigger BS. There is no reporting how much of these "solid cores" are BS, and how much is regular matter # 2. We can be certain that Neptune's BS cannot be bigger than 24,000 km, and compare this to our 1,220 km.

We are interested in BSs power of influence. How powerful is Neptune, and how far extends its power affecting surrounding celestial bodies? Neptune is anchored by the Sun, so is Uranus. Because Neptune affects Uranus; looking at the distance between them, we can ascertain what size of BS how far its influence extends. We have to assume that most of the Neptune's "solid core" is solid BS. So we could examine the influence of 24,000 km diameter BS which affects Uranus some 1,630,000,000 km distance, by the way this is the biggest orbiting distance from the regular orbiting "planets." Now we can say that 24,000 km diameter BS affects other celestial bodies at 1.6 billion km, and the this this rule should apply for the affected planets: the smaller of them – should be more affected. Let us look at our planet; forget 12,756 km actual diameter – the diameter that counts for strength is 1,220 km, because the mantle and the crust are atomic "fluff" which came from this 1,220 km ball. If the 24,000 km BS were white dwarf, the size comes down to 24 km, which could have devastating effects on the Earth from the distance of 1.6 billion km! How something as big as 24 km could be seen from a distance of more than 10 AU? But it has to be found! What if this brown dwarf is 120, 000 km diameter, and its BS is 100,000 km? We do not have table of size related to distance of influence, because we still believe in Newton's mathematical fairy tales!

BIG UNIVERSAL AND LOCAL EXPLOSIONS

These explosions are the biggest in the Universe. From a All of the Biggest Black Spheres (ABBS) "exploding" pieces of them aare flying in all directions, and these a BSs or "compacted matter # 1" which is" BSs matter # 3. We know that BSs are working machines that pump out BBPA. We also know that in supernova explosions these "machines" get "temporary" "broken" (compacted) into neutron stars and white dwarfs. And here is the question: If in a supernova explosion the BS gets "compacted," then in this explosion of the BBS, which is the biggest in the Universe, doesn't the same thing happening as welllso? Let us say that all BS get "compacted" in this explosion; then we should have initially galaxies made of neutron stars and white dwarf only, which "should" "recover" later," and become bran new BSs.

CURENT BELIEVES ABOUT STAR FORMATION AND STAR STRUCTURE

Scientists like the phrase "star formation," and like to repeat it quite often. This is one huge lie – complete and total nonsense. This is their version: Stars are full of hydrogen and nuclear reaction in the center is taking place making helium, and that is the source of their energy. Later on when all the hydrogen supply gets exhausted guess what happens - helium begins to "burn" (burn mean nuclear reaction). Now you should know that for the hydrogen to begin its nuclear reaction temperature of 11.7 million degrees Celsius is necessary. This high temperature comes mysteriously from nowhere. Now the scientific fairy tale requires 100 million degrees Kevin (Kevin is the same as Celsius minus 272 degree) for the helium 'sto "burning." I am reading their minds: "We are scientists, and in order to keep our jobs, salaries and perks coming we have to explain everything about the stars, regardless that we do not know anything about it."

The fairy tales continue: As a result of this nuclear reaction from helium are produced carbon and oxygen is their information, when the . Ffact is that our black sphere inside the Earth creates oxygen and carbon without these fantastic 100, 000,000 degrees, but it would be too boring, isn't it? Our BS definitely produces the first 30 chemical elements- definitely IRON, but I was interested about gold as well?iron also. Black spheres from the smallest to the size of 10 to 20 Earth's sizes produces and "tolerates" iron on top of their surfaces. Bigger sizes black spheres no longer tolerate metals and for that matter any solid substances composed of atom and molecules. As the brown dwarfs get bigger and bigger, only some "fog" remains around their black spheres. The conclusion is inevitable: Scientists do not know anything about stars!

The gases coming from the UNIVERSAL volcanoes on the Earth show what kind of elements our BS is capable of producing. (Do not forget that here on Earth we have except universal, also subduction volcanoes, which are not of interest in this investigation). The gases coming out of the universal volcanoes are as a direct result of the black sphere activity inside of any planet, or a moon. Earth's BS releases BBPA, and through their free association they produce atoms and molecules, this process produces the constant and high enough temperature to keep the Earth's interior melted, but some compounds get into gaseous state and have to come out. And that is what the gases from the universal volcanoes consist of.

ANY CHEMICAL COMPOUNDS AND ELEMENTS ARE PRODUCED FROM OUR BLACK SPHERE! What the scientists are not aware is that they are produced from free association – a process about which they do not have the foggiest idea at this moment.

The fairy tales and scientific nonsense continues with thermonuclear reaction at center of a star where after the supply of helium is exhausted begins the carbon "burning," temperature requirements went up to 600 million degrees C, and report is that this reaction produces neon, magnesium, oxygen, and helium. The scientists were very kind to inform us that this 600 million degree temperature came as result of "GRAVITATIONAL COMPRESSION" - Newton's miraculous "apple gravity" – doing miracles again! Reality check, oxygen is produced by the black spheres in planets and moons in abundance without 600,000,000 degrees K – reportedly Io's volcanic gases are predominantly sulfur dioxide, and the same like here on Earth. Where from came this oxygen and sulfur if not produced by the BBPA? Where from would come iron around the BS if not produced by the BBPA?

The temperature requirements are getting higher, guess what temperature is required at the core of a star in order to have "oxygen burning?" -1.5 BILLION degrees K. So at this temperature "one can make nuclear bombs" out of oxygen. "The principal product of oxygen burning is sulfur." Exactly to SULFUR I wanted to get to – around Earth's volcanoes there is smell of sulfur. Workers are mining sulfur at the vents of the volcanoes. We can be 100% sure sulfur is coming out of all Universal Volcanoes. The reports that Io's volcanoes are releasing sulfur dioxide and the abundance of sulfur from Earth's volcanoes is absolute proof that sulfur is an element that is very easily self-produced, and this is one more proof of validity of my theory, and that is the last nail in the coffin of these nonsensical "star formation" out of hydrogen, and stars being powered by hydrogen as well.

GALAXIES

All galaxies have a central black sphere, but the scientists say that only some have it. Without a Central Black Sphere there cannot be a formation of any galaxy. For that matter any cluster of stars should have a big BS in the center of it as well, and from there it commands all the stars in the cluster. Yes, there is "influence" between all black spheres, stars, and all other celestial bodies, but by all means is not the Newton's gravity! Today's scientists believe so much in Newton's gravity that they do not seek to discover the true nature of this influence between the celestial bodies. Let us not forget that each celestial body is created initially by a black sphere. We have to understand that when one BS has given all of its

BBPA out like in Mars and our Moon we have the process of "transformation and transfiguration" where all of the BS matter # 1 was transferred to matter # 2. has become from "matter # 1" to "matter # 2." In these two matters the indestructible BBPA remain the same, but because now they become atoms in "matter # 2" the density changes. If we assume that the Earth's BS did not change its size from the time it become matter # 1, we can calculate its size and density 13.8 billion years ago., we can calculate what its initial density was 13.7 billion years ago. If after the Big Universal and Local Explosion the Earth's BS was compacted to matter # 3; today's 12,756 km. diameter was only the size of the "naked" BS which was 1,220 divided by 1,000 = 1.2 meter diameter. The Moon was 34 centimeters. In order to know what was the density of the Earth's BS when from compacted become its normal size of 1,220 km put all of today's matter inside this ball. All we have to do is put all the current mass of the Earth inside this 1,220 km diameter ball. We also have to keep in mind that this initial density diminishes through time, and at this time is somewhere around 2,365 tons per cubic meter (calculation done by hand and rounded up twice). The "commanding influence" by all spherical bodies is property of the BBPA. The Earth's current size of 12,756 km diameter has come from its BS 1,220 km diameter, and if it was white dwarf 1.2 meter. ITS INFLUENTIAL POWER STAYS THE SAME AT ANY SIZE, BECAUSE IS PROPERTY OF THE BBPA, AND THEIR COUNT STAYS THE SAME THROUGH ALL SIZES OF THE EARTH. That means real small BS of 1,220 km diameter has the power of a planet 12,756 km diameter.

Keep in mind that we are talking about the strength of our BS 13.7 billion years ago and the strength of our planet without its BS inside. With the gas giants we have a bigger confusion. Let us get for example Jupiter. It has a lot of gas; then it has matter number two of rocky material, iron, and finally at the center is its BS. The scientist are giving the size of some "solid core" of 10 times the mass of the Earth, where I cannot ascertain how much is the rocky material, and the most importantly information how big its BS is? Here is a quote from "Smithsonian" 2020 page 178: "Deep inside, at a depth of about 60,000 km, is a solid core of rock, metal, and hydrogen compounds." We should not blame them for this vague information, because they are not aware yet about the importance of the BS inside, but the size of the BS determines the its strength. That means that an accurate assessment about which of the 4 gas is the most powerful cannot be made until we know the sizes of their black spheres are known.

Galaxies live forever, which means that they would be in the sky forever, but they are constantly in the process of recycling. No galaxy lives longer than 20 billion years because that is probably the length of one "local and universal cycle." of a cluster of galaxies. How this never-ending-; one-and-the-same "nuclear" reaction of creation and subsequent destruction- keeps on going forever and "dynamic existence" of the basic building blocks of the atom keeps repeating itself without losing any matter or energy is a s the biggest mystery ever? Could we say:

"The 'dead' matter is 'alive' forever, therefore IMMORTAL?" That is the inevitable conclusion right now. All galaxies in a particular cluster go through the same "evolution" together. This evolution begins with the Big Universal and Local Explosion, and ends at the next one. This is the Universal Cycle which at this time tentatively I would give 20 billion years in duration, but its real length is to be determined. Today's scientist when talking about galaxies "evolution", they are referring to some "evolution" which supposedly began from some fictional Big Bang, and we do not know how the fairy tale of "their" "evolution" would end. May be the Little Riding Red Hood might tell us.

Here is where things stand right now: There was no "Big Bang" ever; the universe is not expanding; the stars are not made from gas; the universe has existed forever; the black sphere substance lives forever; the universe lives through a dynamic existence of constant repetition of the "universal cycle"; black spheres go through constant transformation and transfiguration in their dynamic existence.

──We live in a spiral galaxy, and might be that only spirals are accommodating for possible life. Overall the universe is hostile toward existence of life. Our Galaxy has in it a lot of planets, so it is accommodating of the existence of "matter number two" and life.

___It has been reported that gas, dust, and planets do not exist in some galaxies. That agrees with my theory. It has been reported that 30% of all galaxies are spiral. Let us put this number at 33.5% and let us assume this scenario: After the Big Universal Explosions in a particular cluster of galaxies, the first galaxies that are created are not the spiral once, but some other kind which later on become spiral. Let us not forget that at the initial explosion there is strong possibility that potential future BSs are "compacted" as neutron stars and "white dwarfs," and need some time for a "recovery." The first created galaxies exist for some period of 6 billion years which might be 3~~0~~3% of the time, then the same galaxies become spiral and live for 33~~0~~% of the time, and finally all galaxies in the local group become one huge ~~a cluster become e~~elliptical galaxy, which with other such galaxies becomes a quasar, and this takes the last 33% of the time to complete the cycle. ~~which congregate for the next big universal explosion for the rest 40% of the time remaining in the cycle.~~ Keep in mind that these percentages are only approximate. ~~Looking at some pictures it is apparent that spirals congregate for an "enlargement" becoming part of huge elliptical, which in its part becoming quasars.~~ The fact that we see different galaxies might be as a result of their evolution in their "cycle" of appearance. ~~Most definitely Q~~quasars are the last transfiguration before the creation~~"preparers"~~ of the biggest black spheres (BBS) ready for the next~~and final~~ ~~"quality control" of its consistency prior of the~~ Big Universal Explosion, which would begin next universal cycle.

~~Here is the scenario: After the quasar, the Big Explosion occurs, and the new cycle begins, but here is a question: because after let's say 20 billion years at the end of that cycle all of the material has to be returned for the next Big Explosion,~~

108

In this universal "cycle" the sequence of events looks like pre-programmed, and I strongly believe in this. Sorry to say this, but the elimination of all matter number two, which is planets, moons, gas, dust, and us is in its initial stages right now– for us it is all matter of luck. We might exist another 500,000,000 years or might not. There are all kinds of hazards – flying huge stones might hit us; supernova explosions near by; the Sun might run out of fuel; our black sphere inside the Earth running out of fuel; big brawn dwarfs or small "white dwarfs" coming from nowhere and removing our BS off its base. The last thing is happening right now, and as a result of it we have: climate change, forest fires, food shortages, atmosphere leakage, global warming, you name it.

TYPES OF GALAXIES

According to current classification of galaxies there are roughly four categories of galaxies: irregular, elliptical, lentil, and spiral. I object to the term "irregular." These galaxies are in a process which is scheduled as a part of the Grand Universal Cycle. They must "collide" in order to congregate and form elliptical galaxy. The Small and the Large Magellanic Cloud are next to our Galaxy according to the script of the Grand Universal Design. Another words, it is time for the local "crunch." Scientists interpret this firstly as a "random" collision between the galaxies, and secondly they talk about some "cannibalism" as supposedly the larger galaxies like to take the "material" of the small once and in the process become even bigger. This is part of their fairy tales!

To recapitulate the history and the future of our Galaxy: After the previous Big Explosion that created the Local Group our Galaxy went through lenticular look, now is it late stage of being spiral one, and let us say that life is possible only in spiral galaxies. The new huge galaxy that would be created from the material of our "local group" of galaxies would become one elliptical, and supposedly without "matter # 2" present in it any more. All gas, dust, planets, atoms and molecules would be disassembled to its basic building particles of the atom and "stashed" back to where they came from in first place – inside of the black spheres. So as you can see that the basic building particles of the atom are UNDISTRUCTABLE AND "LIVE" FOREVER. Now, I am trying to discover how and why they "live" forever? Obviously this is a perpetual motion, but what is the "mechanism" that makes this "perpetual motion" go indefinitely? At this moment, considering the fact that the energy in the universe is free, I imagine that this BBPA are forever unstable and have to go through this constant repetition of this never-ending "cycle". The conclusion is that the BBPA have entered in some self-propelled cycle to "live" forever absolutely the same life. Nothing of this kind can ever be achieved on Earth! All chemical or atomic reactions here come to an "expected end," so we do not have an analog of what is happening in the universe in grand scale – everlasting, never-changing, cyclical NUCLEAR REACTION. The smallest building particles of the atom relive forever the same absolutely-not-changing (how to call it - life; performance; reaction?). This is a new-dimensional phenomenon, which requires new way of thinking and terminology. We could say that the BBPA are "alive," because they never die. There is no "logic" in the whole process, but the energy is FREE FOREVER inexhaustible without loosing the smallest amount. My suspicion is that ENERGY IS NEVER SEPARATED FROM THE PARTICLES THAT CARRY IT. This cycle is like a living entity that never had a "past" where it "began," and can never have a "future" in which thing would come to an eventual "end." All the changes that are "observable" are into the Cycle, and if we "disregard" their constant repetition – there is no "change" "ever!"

Who are we, and how we fit into this larger picture? We are somewhat "thinking" and might have some "cognizance" as this cognizance is variable. We are very fragile as creatures inside and outside. This book deals with our "fragile" existence where "ultimate" our lives depend 100% on external favorable conditions "delivered" to us by some black spheres. We have two "mothers:" One that brings us to life, and a second one which is the ever-present and favorable to us but fragile. Space aliens can go through the Galaxy and look for planets like ours, but we are not that developed mentally and technologically.

We have to know more about the universe. The more we know; the better are our chances of "survival" without the favorable condition we have on Earth. We could try to "imagine" and "prepare" for this "survival." Imagine that we have created "self-sustaining" colony on the Moon for the purpose of "survival" of the human race in case if something wrong happens to the Earth, and it is happening right now. What a gloomy story! Imagine all the people on the Earth are being killed. On the Moon remain the "survivors" from the human race. The Earth is uninhabitable. The people on the Moon are "space aliens" already. They have to find some good place like the Earth to live "normal" lives. They do not know in which direction to go, as they don't know of any living planet. Let someone else continues this story... We all collectively have to write this story, because it is our collective life. Let me say something crazy: May be instead of eating food, we should take our energy from electricity? Of course these ideas are not for this book. ~~story...~~

~~What never die are the basic building particles of the atom, and the Moon is made out of them. So that the Moon's BS has transformed itself to atoms and molecules from which the Moon is made now.~~

GALAXIES CLUSTERS

"Abell 2218" is made of a cluster of huge elliptical galaxies packed in very small space allegedly 2 billion light years away. This particular cluster plays well into my theory. It consists only of very large elliptical galaxies which are o~~i~~n their way to get together and form one or several of the Biggest BS for the "Big Universal and Local Explosion" (BULE) for the beginning of the next Universal Cycle. When the~~y eventually get together~~ ~~eventual enlargement gets to the certain limit; they would become~~ ~~, the~~a newly created ~~huge galaxy becomes~~ quasar. I suspect that the state of quasar is the last "purification and preparation" of QUALITY Black Sphere Substance (BSS) before the next BULE.~~, after which the flying pieces of BSS has to be of high quality in order to perform their duties of "working" BS as suppose to.~~

——Our local cluster is on its way to become like one of the elliptical galaxies from Abell 2218 ~~concentrating all the BSS in one place, and in the process destroying our world of "matter number two" consisting of atoms and molecules.~~

THE BIG UNIVERSAL AND LOCAL EXPLOSIONS

The Big Universal and Local Explosions (BULE) are the biggest explosions in the universe – there are no bigger explosions than these. The "local" such explosion that happen 13.8 billion years ago in our neighborhood has been ~~mistakenly~~ named "Big Bang", and mistakenly has been assumed that from this explosion somehow ~~was created~~ the entire universe was created. Many other wrong~~other mistaken~~ assumptions have been made in the past – the important thing is to correct them and to forge ahead, and that is what I am doing. To ascertain the size of the ~~At this moment I am not quite certain of the size of the~~ Biggest Black

Spheres is difficult. ~~Possible (BBSP) right before the BULE.~~ Let us say that 200 new galaxies would be created from this explosion. This number is arbitrary, but ~~W~~what we could be ~~we could be~~ certain ~~of~~ is that from the previous BBPA none is missing.~~at least 40 galaxies are packed in one of these BBBS.~~

~~As we know already in the supernova explosions the BSs get "compacted" and "temporary" "damaged" (no longer capable of releasing BBPA due to internal structural "damage"). The question is: In this biggest possible explosion are the BS "damaged" in the same way? Let us not forget that they are always capable of "recovery." We see this happening when neutron star sucks up the "staff" from another star, and in doing so they "recover."~~

~~BBPA IN BLACK SPHERES LIVE FOREVER~~EMPTY SPACE BETWEEN CLUSTERS OF GALAXIES

We already know that black spheres substance lives forever in this never-ending and never-changing UNIVERSAL CYCLE. When one cycle ends it is followed immediately by another one which is absolutely the same. All galaxies in a cluster ~~or a "supper cluster,"~~ go through this cycle TOGETHER, otherwise there would be constant enormous explosions through a cluster and disruption of the "preordained" ~~evolutionary~~ process. The cycle is performed at different times in different clusters of galaxies in different parts of the universe. The enormous empty spaces between clusters of galaxies can be explained with the DISRAPTIVE EFFECT the big universal explosions have on surrounding clusters. If clusters are right next to each other, the explosions in one cluster would affect the neighboring one. The definitions of "cluster" and "super-clusters" should be put under review!

HOW TO OBSERVE~~WHY SCIENTISTS NO LONGER CAN SEE~~ THE AND DISCOVER THE "SCIENTIFIC TRUTH?"

The life of the whole universe consists of these constantly repeating cycles in different locations throughout the universe. That means that all the different details of the same cycle are constantly going on in front of our observational tools. All we have to do is to observe them, and put different parts of the~~is~~ puzzle together. This book is giving you the rules of the puzzle. Why the scientists although doing constant "investigations" would never come to the right "conclusion?" Because in first place they do not have some "guidance" as to where the right conclusion might be, and second with their 85~~90~~% brain-washing believes; they are depriving themselves from "seeking" and "finding" the "scientific truth." Right now scientists are so much prepossessed with "Big Bang," "expanding universe," "Doppler shift," "Newton's gravity," "Sun filled with gas," "the ~~-~~speed of light is the speed limit of the universe," that they are almost "blind~~,~~;" and ~~in order~~unable to carry a real investigation. Examples, because the Sun supposedly is filled with gas ~~which is not true~~, it should not has surface, but the surface of the Sun is clearly visible in this picture from August 1973 where the "black" curvature of the Sun on top of the picture is clearly visible. And that brings another question: Should theoretical assumptions take priority over observations? The answer is NO! In 2002 two American teams armed with the latest investigating gadgets came to the conclusion that universe is not only expanding, but is expanding even faster than previously assumed. How they came to the conclusion that the universe is expanding even faster when is not expanding at all? The answer is: That is what they were looking for!

The universe has existed forever, and is not expanding at all, but who needs expanding universe – the "Big Bang," the "Doppler shift," and "Hubble's constant of expansion," which are all wrong. One of the biggest nonsense - Newton's "falling apples gravity" is widely used to "explain" even "eruptions" of volcanoes on Io ~~(Jupiter gravity pulling on one side, and the moon Europe pulling on the other side)~~; creation of stars and planets from gas and dust right now; creating fantastic temperatures from zero to 12,000,000 degrees C (Temperature required for the initial formation of a star.); non-existing gravity making planets go around the Sun and moons around planets – ALL THIS IS ABSOLUTE NONSENSE ~~all this is all absolute nonsense~~. Because Newton's wrong formulas and assumptions "more matter is needed" otherwise the stars might fly out of the galaxies. To patch up Newton's nonsense; they invented another one – "dark matter." ~~ow what should explain the fact that there is not enough mass in galaxies to hold the stars inside them together? Keep in mind that these people are not quite clear in their definition of "mass" – the mass they give for the Sun is absolutely 100% wrong. Here is their "solution" – thinking-up invisible "dark matter" otherwise Newton's gravity obviously cannot hold stars in galaxies together, because is valid only for falling objects to the surface of celestial bodies.~~ Doppler Shift is another's complete nonsense. ~~that light from a star or galaxy or portion of it gets some sort of "red shift" because is going away from the observer leads to the insane "conclusion" that some galaxies that are so far from us, that we cannot see them, actually move away from us at speeds higher than the speed of light – how they came to this preposterous conclusion, through "phony" mathematics (all this nonsense are backed up with mathematical formulas). Obviously the Newton's fake gravity cannot hold the stars inside galaxies – solution, how about inventing some bogus "dark matter."~~ Are these people sane~~, or crooks, or cannot think logically~~?

~~We can say that the life of the universe is the life of the black spheres. I would say that black spheres "live" DYNAMIC "life." They exist in a "perpetual motion", but what makes it "tick?"~~ The BBPA "live" in perpetual motion. We cannot achieve "perpetual motion" here on Earth, because we do not have access to free energy, which is one of the differences between our world of "~~already made~~ atoms and molecules" and the 98 7% of the real universe ~~that is made of BSS~~. Our "world" is fundamentally different than the physical world of black spheres, so the notion that what we discover in our labs is valid throughout the universe is almost 100% inaccurate. What we discover in our labs is valid only for "matter number two" which is atoms, molecules, and other particles, but never black sphere substance. Something we cannot achieve on the Earth. Black spheres have the ability to do things without the waste of any energy. We are not aware of any live entities that are basically made of the basic building particles of the atom, so we have to be very careful of our descriptions and assumptions. I would say that this is some sort of "runaway" self-supporting nuclear reaction. Most definitely, one of the most important feature in trying to understand what makes this perpetual motion going on forever is the availability of free energy. So we have free energy, and some cycle of black sphere substance of constant transformation and transfiguration. Matter number one is never lost in the most infinitesimal amount, and "matter number one" never becomes energy and vice versa. We know only of biological creatures that are alive, welcome to the real world – the universe is one giant LIVING NUCLEAR REAVREACTION. Now, let us not be in a hurry and assume that these black spheres are simply "packed" with basic building particles of the atom, because they act as if they are ALIVE, but we have to take this into consideration in order to have a clear picture of these phenomena. The term "alive" should be understood as this constant repetition of the Universal Cycle. The immediate and inevitable question is why REPEATING this cycle forever, and let's not rush into fast and inaccurate conclusions, which seems to me is our "modus operandi." The repetition of this cycle is dictated by "internal" needs of black sphere substance, which constantly has to go through these "changes."

Many erroneous assumptions have been made in the past, but the march of acquiring true knowledge by the human race is ordained by God, therefore has to happen. Satan might be workings through people like Einstein, Hawking, Doppler, Eddington, Hubble, Newton, and many others. I am not sure is stupidity on their part and gullibility also tools in the hands of Satan? Examples of simple "suggestions" by Eddington (that stars are filled with gas) and Doppler's "red shift" were turned to "laws" of the universe, and today the human race comfortably believes this absolute nonsense. The struggle between good and evil will never end. The spreaders of half-truths and scientific inaccuracies are not after the scientific truth, but after the good jobs, salaries, perks, and fame. And all these fake historical "discoveries" are backed by "mathematical" formulas, theorems, and wrong physics. What that tells you about physics and mathematics? I would tell you what they are – convenient tools in the hands of the DECEIVERS – in the hands of the servants of Satan. Now the "supper" computers are joining in spreading "scientific" nonsense, and wait for the "artificial intelligence" to join the chorus.

As you can see in my version of events - "stuff" from ~~40~~200 galaxies is used for the making of 200~~40~~ new galaxies, and every atom has been accounted for – matter is never lost in the universe in the smallest amount. It has been reported that that the mass of two hydrogen atoms is bigger than the one atom of helium created. Advice to the scientists, please look for the missing mass, because mass is indestructible and Einstein's formula e = mc2 is fake.

Let us see now, how the rival "big bang" would compare~~do~~? "Out of nothing, came everything." The so-called nothing is something mythical and as big as a walnut, and the "everything" is~~are~~ all the countless galaxies and stars. Are you kidding me? In order to believe this one has to be severely mentally impaired. Are all people mentally impaired in the world? Of course not – they rely on the scientists for answers, but the scientists are mired in nonsense like Big Bang and many others. Would that mean that from no mass came all the mass in the universe? "Before the big bang, there was no space and time." The universe found itself without time and space before the big bang, but where were the Little Riding Red Hood and Pinocchio at this time – no one mentions it. We are definitely smarter than monkeys, but are we smart enough to discover the universe? By the way, Arthur Eddington in 1930 was informing us how many protons and neutrons are in the entire universe, without knowing how big it is! Is this person normal? And on top of it being a SCIENTIST~~calling him a scientist~~? He has said that the temperature in the center of the stars must be higher. He is "smart" with the sign of minus in front of it – therefore STUPID. No wonder he had a friend who was border-line idiot, but he made it big – real big – the mathematical crook Einstein, and Eddington admire him for that. Einstein came with the idea of "singularity" in the middle of black holes, where space becomes time, and time becomes space, AND THE IDIOTS BECOME THEORETICAL PHYSICISTS ~~but I guess the idiots become theoretical physicists~~.

All black spheres are made from the same substance. If you have played with the metal mercury, and separate small pieces of it; they would become like little spheres, but when putting them back together they merge into the bigger piece flawlessly~~;~~. BSs are like that. ~~The supernova explosions are from material accumulated on top of it, and are probably thermonuclear as enormous amount of hydrogen accumulates around the star, and all it needs to trigger the explosion is high enough temperature~~

~~Stars do not tolerate matter of the second kind on top of them. Reason is simple – the so-called "solar wind" is actually fast expulsion of BBPA which creates bigger particles, atoms and molecules, and in this process heat is generated from the free energy.~~ The bigger the star, the more BBPA are expelled per unit of time, and the tolerance of accumulation of any material ~~of any kind~~ diminishes exponentially. For example, the not-too-f~~urious~~ast expulsion of BBPA from our Sun – allows the existence if this "photo sphere" which in bigger stars is not "allowed". As matter # 2 ~~of the second kind piles up on top of~~ accumulates around

115

them, and the release of BBPA continues relentlessly, a mixture of explosive elements are piling, and when the temperature gets into millions of degrees, an explosion is inevitable, and hydrogen becoming helium. After the explosion we see the star, so this explosion is not from inside. When biggest black spheres explode, the story is different – the new universal cycle begins with the creation of new galaxies – made of the exploding pieces of the Biggest Black Sphere. Around different stars there are different elements; this means that the speed of BBPA release FAVORS this process.

Instead of saying that black spheres live forever, it is more accurate to say that BBPALACK SPHERE SUBSTANCE (BSS) lives forever, because separate black spheres some times merge together, or "disappear" when creating SECOND MATTER which is atoms, planets, moons, and humans among other things. What remains forever is the BBPASS. This is the primary matter of the universe that lives forever. All stars are BSs, therefore are filled with BBPAlack Sphere Substance (BSS). In the spiral galaxies where from about 23% of matter # 1 number one isbecomes converted temporarily into matter # 2 number two,; where life is possible. for us. If we say that the spiral galaxies exist 30% of the time in the cycle, and are 30% of all galaxies, then in each cycle of 20 billion years, fFrogs and humans have the chance to exist in this window of opportunity which is about 5-6 billion years. UnfortunatelyUnfortunately, smarter animals like us develop last, and even with space aliensalien's interference we are still not real humans. So, the space aliens, who are the Gods as well, did some genettic engineering on us, and at this time are giving us some scientific knowledge when we sleep, but in order to have some real progress – we have to do our part in the grand design of things as well. My assessment is that we are slow learners, but the Gods' are not giving-up on us. Let us hope that the Gods know what they are doing, because the evidences are that we live in scenario already written by them, and we CONSTANTLY FAIL TO LEARN OUR LESSONS! for us. The Earth is not always accommodating for animal existence. Basically, they do not want us to die like the dinosaurs, that did not know what hit them, but with today's scientists' myopia and gross scientific misunderstandings, things are not looking promising, so you should not disregard my findings.

ALBERT EISTEIN

I have an enormous dilemma, a task that is beyond anything. I am certain that Einstein is an absolute idiot, yet the world is convinced that he is some sort of genius. First I want to tell you how I reached this conclusion. At age 65 I read Einstein, and up to that point I accepted him like everybody else. As I was reading his booklet, it struck me how stupid he was, and also the fact that he did not want to discover the universe, rather he wanted to impose on the universe his twisted imagination, but there is more to that – he wanted TO SHOCK PEOPLE, so he read other scientists like Lorenz in order to find something "unusual" and "shocking." He is an idiot and all his theories of relativity are "figments of his sick imagination;" rather than some sort of "discoveries." We also have to reexamine some of OUR "acceptance" of scientific "discoveries." There are different fields of specialized knowledge and it is only logical that we have to turn to the "specialists" for guidance and advice. THE HUMAN RACE NEEDS THE "SCIENTIFIC TRUTH," BUT FOR CENTURIES HAS BEEN MISLEADED IN THE FIELD OF ASTRONOMY. This is a difficult subject that relies on many assumptions and stipulations. I am ready to go through any book or textbooks with math and physics that defends the current misconceptions, and prove them all wrong. Reading through some textbooks, I could see the accumulation of scientific nonsense throughout the centuries.

Let us look at ourselves. The general public and the scientists have some preconceived notions for ACCEPTING DISCOVERIES. One of them Is "predictability," so when Einstein stated that gravity bends light, and in the solar eclipse in the year of 1919 this was seen, the door for more "nonsense" on his part was wide open, and sure enough the "mathematical" crook "milk" this opportunity to the greatest extent possible to the rest of his life. On the issue of the "gravity bending the light," which scientists still believe, I would say, this is not true. So how it cannot be true, when it is obvious to see? The electromagnetic field bends light not the gravity, and because both occupy the same space mistakes in judgment do happen. Also the atmosphere plays a role. The atmosphere is like conduit through which light images from behind the Sun can be seen on the side.

"In 1905, Albert Einstein rejected the idea that there is any absolute or 'preferred' frame of reference in the universe." from p. 40 "Smithsonian Universe" 2020. Let's analyze this sentence. It implies that Einstein after some "rational thinking" which he is incapable of doing, the great scientific genius "thoughtfully" decided to "reject" whatever it is. So, we have a case of accepting an idiot as a genius. Einstein is a "mathematical crook" – he used his mathematical skills to lie to the gullible public, and constantly inventing hard-to-believe astronomical nonsense to keep his "genius" image alive. He was a pathological scientific deceiver; he knew what he was doing; and all worked for him perfectly, but not for the public IT IS A DECEPTION! There is another aspect here; all the scientists today DO NOT KNOW that the gravity cannot bend light; gravity do not warp the "fabric of space" nonsense; or that in the center of black spheres there is some sort of "singularity" which is absolute "scientific" garbage. The problem with Einstein's "descriptions" is that they do not lead to or encourage any farther discoveries, because they are dead-end "nonsensical baloney." Here is how the "naïve" human race is helping these "scientific deceptions" to happen: first they label someone "genius," then whatever nonsense he say, becomes some sort of "law" of the universe, and at this moment 85~~90~~% of the "science" of Astronomy is pure "scientific garbage." There is also the phrase "common sense," which is absolute misnomer. There is no such thing as "common sense." If someone describe to several people something strange and then asks the listeners one by one to explain the same thing in detail, you are going to hear many "different versions." There is no person who has done more DAMAGE to the science of Astronomy than Mr. Einstein. He received Nobel Price for his lucky guess that photons are acting simultaneously as particles and waves - most likely he stole this idea from somebody else – that is what he was doing his entire life. He was a very poor person as a young man working as a mathematician, but he LEARNED his lesson fast, and become a "mathematical crook"~~,~~ and ~~became~~ "theoretical astrophysicist," but the precise definition of this profession should be "professional liar." ~~His first lucky lie that got him in the road of fame was that "gravity bends light."~~ Neither Newton nor Einstein understood the real "gravity" that affects galaxies, stars, planets, and moons ~~in the universe~~. Einstein was the ~~angler~~ANGLER, and his fish were other scientists and the general public. Since then he kept his image of some sort of "genius" who "knew" the universe and "churn" lie after lie. His half-cooked lies still stand TODAY! ~~The dumb moron has fooled you people, wake up!!!~~

    Both of his theories of relativity are ABSOLUT AND COMPLETE NONSENSE. Let's start with the "time." There is only one time throughout the universe. The only problem is that planets, moons, and stars are spread throughout the space in big distances. The time distance between the Earth and the Moon is 1.3 second, which means that if a light is flashed on the Earth, after 1.3 sec would be seen on the Moon. I was thinking about this time phenomena, and reached a conclusion – you are the center of your time. Your time stays with you no matter where you go, so is the time of our Earth, no matter where on Earth one can be there is local time which is coordinated with the time throughout the entire planet. Recently I watched one TV show defending Einstein's "time" nonsense. One atomic clock was left on lower portion in the state of Arizona, and a second one was placed overnight at some higher ground in same state. Next day, the clock from the higher ground was hooked to a computer, and showed some 20,000,000 of a second difference with the other clock, and this suppose to be proof of Einstein's "time." You see, right now, I am in the role of a scientific investigator. Because I cannot verify any of this experiment, I have to call it "scientific propaganda" in favor of Einstein. I am sick and tired of these broken clocks even if they are "atomic". This is not isolated case of blunt "scientific" propaganda. This "propaganda" is based on STUPIDITY; MIOPIA; INABILITY TO PERFORM SIMPLE TASKS; SCIENTIFIC MAFIA COMPLIENCE.

    At the end of the last century I guess two US teams were tasked to determine what is the true rate of expansion of the universe? For a long ~~time~~time, the complete nonsense called "Hubble's constant" was suspected to be to~~o~~ low of a number (this is a constant for the rate with which the universe is expanding). In 2002 both teams came to the same CONCLUSION - the universe is not only expanding, but is expanding even faster than previously assumed!!! The fact is that the universe is NOT EXPANDING AT ALL, so how these two teams using the latest gadgets of x-rays and satellites came to this conclusion? The answer is simple - that's what they were looking for, and that is what the scientific mafia expected them to find out. So, all the scientists are on the "right" bandwagon. At this moment the scientists are so well assured that they know the "scientific truth," that they even hide their "discoveries" which are basically "discovery of nonsense." I am trying to fathom this mass-scientific-herd mentality. First if one person is to become a scientist he has to be "ONE OF US," which can be achieved through the "right" connections. It is not important how good scientist one is, but how well politically he is connected, therefore the "scientific knowledge" is decided by the "right political connections." The direction of the scientific field is decided by people who don't know or understand science, and if this is not a "scientific mafia," I don't know what is it? The main thing is that real science goes out of the door. Then there ~~are~~is the money to be made, the good carriers to be had, and the good life with it. Have you heard of a professor teaching something else than the mass deceptions mandated by the scientific mafia?

TIME IN THE UNIVERSE

I feel that it is below my dignity to address nonsense like this, but here it is. Scientists that we have to respect as they are on top of their field accept Einstein's idiotic claim that there is some difference in measuring time in different situations and locations. Before Einstein, people thought that there is only one unifying time throughout the universe, and that was correct, but the genius-idiot appear on the scene, and among other nonsense said that time differs in different locations. More than that – he ~~claim~~claims that time is an illusion. I am trying to "fathom" the sewer of his thought process, but people prefer to believe in his nonsense over the reality. I guess it is more interesting and more mysterious, wouldn't be boring if the time was the same throughout the universe, well let's believe Einstein and make our lives more interesting, shall we?

There is only one measurement of time throughout the universe whether it is inside black spheres, strong gravities, or high speeds. Let us assume that you are going in a space trip equipped with a quality clock. Your clock continues to function normally throughout this trip, if it is made that way. Upon return your clock has to show the time laps as if you have never left our planet. All these time dilations are as a result of improperly made devices. If you went to the Moon and from there flashed a light toward the Earth, and clocked it at certain time according to your time device, it will arrive to the Earth at the same time plus 1.3 sec (the speed of light). ~~How to synchronize the time between the Earth and the Moon?~~

~~Technically we cannot, but the Moon's orbit is elliptical, which means that the distance between them varies. I bet that all the space agencies in the world know that Einstein's "different time" is nonsense as they track their satellites.~~

SPEED OF LIGHT IS NOT THE SPEED LIMIT OF THE UNIVERESE

Quotation from a textbook~~:~~ ": "The speed of light in empty space is one of the most important numbers in modern physical science. This value appears in many equations that describe atoms, gravity, electrical theory, and magnetism." Because the speed of light is not the speed limit of the universe, all the equations that contain it are TRASH! According to Einstein's special theory of relativity, nothing can travel faster than the speed of light." The little punk Einstein based on this assumption his famous formula e = mc2, which is a pure hoax, but he needed the fame and the money, and got it.

Proof against it. In 1987 explosion of a supernova in the Large Magellanic Cloud (a small galaxy in collision course with ours), the neutrinos arrived three hours before the light, and were detected in the US and Japan. The distance neutrinos and the light from this explosion have to travel is about 168,000 light years, and the neutrinos came 3 hour earlier. The fact that neutrinos move with higher speed than the light in water was discovered by a Russian scientist Cerenkov in the year of 1934. The question is why back then this fact was not accepted, and with all the present evidence is still not accepted? Here is my understanding of the different speeds of the electromagnetic radiation spectrum. Every-body knows that the slowest are the radio waves AM, and looks like the fastest should be the gamma-rays, because there is obvious correlation between the length of the waves and their speeds. Light is somewhere in the middle. On top of it, the light consists of several colors, and each color has different wavelength, therefore each color should travel with different speed as well. The difference between their speeds ~~is~~should be quite miniscule and very difficult to detect, yet ~~should~~ exists. How scientists can detect this difference in speed? The longest wavelength has the red light of 700 nm therefore it should be the slowest. The blue light has the shortest wavelength of 400 nm, therefore should arrive first. ~~In order to detect that the blue light arrives before the red one,~~ Tthe scientists should develop~~have~~ special equipment ready to record this phenomenon, and the other requirement is that the light should be coming from really remote source so that some tangible difference in arriving could be possible to detect.

Faster than the light are the x-rays and the fastest should be the Gamma rays. The chart of the electromagnetic spectrum I am looking at do not includes neutrinos about which we know for sure that are faster than the speed of light. Einstein even assumed that the electrons move with the speed of light, uses phrases like "particles in motions" referring to electrons – what a moron.~~.~~ In his writing he states that all colors of the light should arrive simultaneously. The speed of light is exactly that - SPEED OF THE LIGHT and nothing more, and is not the speed limit of anything.

Some sort of faulty experiments have been conducted and somehow erroneously assumed that no matter with what speed different items are moving the light will reach them simultaneously. Then the story goes that if one chases a bullet, eventually can overtake it, but light travels with such speed that nothing can surpass it. Guess what - all this is baloney – if the light travels with a certain speed, and if something else travels faster, then it can chase it, surpass it, or if it is in front of it – the light would never be able to reach it. Humans, grow-up! Enough of these childish conclusions! Could there be conflict of interest? Yes - and serious one! The human race is "feeding" a "scientific monster" that is giving you "scientific nonsense", and that is dangerous for your collective health, life, and survival as species."

STEFEN HAWKING

Stephen Hawking has to be addressed not for any other reason, but because he wrote so many books of astronomical NONSENSE! He suffers from the same type of mental deficiency as his mentor Einstein who left the door wide opened for the "nonsense-talkers" like Hawking. "Mental deficiency" means stupidity, but as a member of the scientific mafia, he received many rewards for being one of the "good boys," and like Einstein he received prestigious jobs like the one in Cambridge. Einstein specialized in "shocking statements" to keep-up the perception that he is some sort of a genius, when Hawkings is like all the contemporary "luminaries" in the field of Astronomy. All these people that write books on the matter, go on television, are professors, or have PHD, have some things in common: Firstly, they CAN memorize what they read; second, they UNDERSTAND what they read; third, they can EXPLAIN to their students what they have read, but what they CANNOT do is THINK or DISCOVER anything what-so-ever. So the real DISCOVERY of the universe was left for one retired unsuccessful building contractor which is I. You want it build, we'll build it; you wanted the universe discovered – it is already done.

GRAVITATIONAL RADIATION

"A gravitational wave is a ripple in the overall geometry of space and time." "From the equations of general relativity, it is possible to prove that gravitational radiation moves outward from its source at the speed of light."

~~what my real calling was - turn out that actually I should have been scientist in the field of Astronomy - all these gravitational waves ripping through the cotton fabric of space and time. I understand it all. All this make sense to me."~~

Comments on these two quotations: Why there cannot be any "gravitational waves?" Firstly, this statement is based on Newton's gravity, and second on the theories of the super-idiot Einstein. If something is proven mathematically, therefore is legitimate and true; goes their reasoning. All of his Astronomical nonsense is mathematically and physically "garnished." The first "mathematical crook" was Newton, and after him the master of deceptions and complete idiot Einstein, but now I have employed myself as a chief of the world scientific police. MATH IS ILLOGICAL AND CANNOT DISCOVER OR PROOF ANYTHING! How many fake theorems are out there; I do not know. Notice th~~is~~ese ~~quote~~words: "from the equations of general relativity." The scientist who wrote this BELIEVES that these EQUATIONS are legitimate! Einstein's mathematics is simply adjustable garbage. Continue looking for "gravitational radiation" if someone ~~is~~ pays ~~ing~~ for it, but, know that no such thing exists.

SUPPER COMPUTERS

They can be made more and more "supper" through adding speed and memory to them, but they cannot become "smarter" than the scientists who are telling the programmers what to put in it. So, again comes down to how much the scientists know in first place, and as I am telling you - they know next to nothing. "Garbage in – garbage out," okay, how about "nonsense in - nonsense out," or "we don't have a clue about the universe, please "fast and smart computers, make some discoveries for us." Computers are fast calculators with a lot of memory, and that's how they are supposed to be used. Notions about artificial intelligence, sounds like~~:~~ "~~:~~ "We know that we are stupid, but now with ARTIFICIAL INTELLIGENCE, we are about to become SMARTER THAN OURSELVES. Caution, do not give this "artificial intelligence" to the monkeys, because they might become smarter than us! … But ~~T~~there are rumors that the monkeys already have gotten hold of the artificial intelligence technology and ~~might~~ have become smarter than us.

WHERE IS THE MISSING MASS THE SCIENTISTS ARE LOOKING FOR?

Why mass is missing in the universe? There are three reasons for that. The first one is Newton's phony gravity. The second reason is the wrong assumption that stars are filled with hydrogen gas, ~~. The last mistake~~and that eliminates 95% of the universe's mass. The Sun's density is given at 1.4 tons per cubic meter, when at this moment, as it is going down is at 2,356 tons per cubic meter. The third reason is that the scientists do not know about black sphere substance, and about its mass and density, and finally they do not have a clue about the real and "commanding influence" of the BS that rules the universe, and not the no-existing gravity.

ISAAC NEWTON

Let's start with Newton's gravity. Isaac Newton, THE GREAT GRAVITATIONAL DECEIVER, made three convenient assumptions, presently labeled~~called~~, "Newton's laws of motion."~~;" in order to peddle his GRAVITATIONAL DECEPTIONS!~~ I would call them: ~~Here is my definition:~~ "Newton's nonsense of motion." His first law is called the Law of inertia~~:~~ ": "A body remains at rest, or is moving in a straight line at a constant speed, unless acted upon by an outside force." These "assumptions" are very "convenient" for his deception that the Sun's "gravity" WHICH APPLIES ONLY FOR FALLING OBJECTS somehow "captured" the planets in the Solar System. I have the feeling that we are in a kindergarten, and we are asking the children to explain the Solar System, and among them is the little Isaac Newton. ~~These "laws of motions" were necessary for this "mathematical crook" in order to create his FAKE GRAVITATION FOR THE UNIVERSE. He NEEDED planets to move with constant speeds in a straight line, so when his no-existing gravity would make them orbit around the Sun.~~ This is not a scientific "discovery", but "scientific deception." Let us not blame only Newton, the scientific community and lay people have accepted this nonsense to this day.

From Kepler's third law and Bode's law we know the planet~~'s~~'s' distances from the Sun and their speeds are PREDETERMINED by the Sun. This is very important to understand, the Sun not only determines the distance and speeds of orbiting planets, but also has extremely strong hold on them. The strength of this "hold" is not explored yet. If a planet is pushed out of its orbital spot; it might be destroyed. The Sun is a star in a system within a bigger system, and creates its own system THROUGH ITS SIZE, so the "ORDERS" of the Sun ARE THE ORDERS OF ITS SIZE – different size; different orders. So how a planet could move~~s~~ with SOME CONSTANT speed before being captured by the Sun, when each planet has DIFFERENT assigned orbital spot and speed by the Sun? Then has to enter in the EQUATORIAL PLANE OF THE SUN? This argument renders his "assumptions" or so-called "laws" absolutely invalid and absolutely nonsensical. I would argue that scientists like Kepler and Bode were the REAL scientists, and since Newton's reliance on mathematics the science of Astronomy turn to the worse, then came the supper-deceiver Einstein, and with today's reliance on SUPPER computers – the scientists CANNOT discover a thing because they DISREGARD the observations and facts in order to support the "party-line's inaccuracies" of the scientific mafia. Here is a quote from "Universe" 4th edition Kaufmann 1994 "Until the mid-seventeen century, virtually all mathematical astronomy was entirely empirical, characterized by trial and error. From Ptolemy to Kepler, essentially the same approach was used. Astronomers would work directly from data and observations, adjusting ideas and calculations until the right answers finally emerged. Isaac Newton introduced a new approach. He made three assumptions..." Here is what I would say: Up to Newton's time - scientists were doing the right thing – observing and adjusting, but Newton came with his phony assumptions and promoted his phony mathematics and gravity, and the world took the bait, and I guess since then we have Einstein's baloney, Doppler's shift nonsense, Hubble's expanding nonsense, and the list goes on and on. Newton's second law says: "The acceleration of an object is proportional to the force acting upon it." Here is the formula $F = ma$. Where F is a force, "m" is the mass, and "a" is the acceleration. Notice in this formula is used so-called MASS, which is one of the biggest inaccuracies and misconceptions about the universe. Today's masses and densities given for the celestial bodies are absolutely inaccurate. The only accurate density measurement might be that of the Earth. For certain masses and densities of Venus, the Sun, all stars, and all gas giants are absolutely INACCURATE. One book says: do not confuse "mass" with "weight," but they are CONVERTIBLE, and mass is measured in kilograms, and weight is measured in pounds or newtons. All these "measurements" have been DERIVED on the surface of the Earth. So, the so-called "universal gravity constant" has nothing to do with the real universe. The deception comes from the assumption that the matter on the Earth is the same throughout the universe, but in reality they are substantially different.

Newton's third law is: "Whenever one body exerts force on the second body, the second body exerts equal and opposite force on the first body." He thought that planets exert gravitational force on the Sun. A joke: If you see the Sun wobble, know that some planets are exerting pull on it. And even today scientists think that Jupiter and the Sun are some binary system where Jupiter is not orbiting exactly the center of the Sun. Ladies and gentlemen, the Sun and Jupiter are made of the same substance. Here is the ratio of their REAL sizes: Sun's around 1,400,000 km diameter; Jupiter's – 23,000 km diameter. These two celestial bodies are in "relationship" in a SYSTEM. This system DICTATES Jupiter to orbit precisely around the CENTER of the Sun!!!

Now hear the shocker: Newton and the entire scientific community since his time DO NOT UNDERSTAND THE KEPPLER'S THIRD LAW, AND AS A RESULT OF IT, THE CALCULATIONS FOR THE SUN'S MASS ARE ABSOLUTELLY WRONG. Kepler's third law is very simple. It states that for the Earth this equation to be valid: "p" squared should equal "a" cubed, where "p" is the time of one full orbit around the Sun and "a" is distance from the Earth to the Sun. NOW PAY ATTENTION IF YOU ARE TO UNDERSTAND THIS COLOSSAL MISTAKE. By the way, Kepler's third law is good one, but works only if one CONDITION is met. If this condition is not met, this law is invalid, and that is what Newton did! What is this condition that is so important? Both measurements for the planet Earth which are the orbital distance and the time of one full ~~orbitcircle~~ orbit around the Sun has to be taken conditionally as 1.0 (one). The distance from the Earth to the Sun happens to be 150,000,000 km, but we must take it conditionally only as 1.0 (one), because only 1 multiplied by itself three times would equal 1. Newton did not have the brain power to understand this! ~~Same conditional N~~number 1.0 applies also for the time of one complete orbit~~al rotation~~. This mistake was made in 1687; that means that for more than 300 years no scientist, mathematician, or laymen discovered this gross mistake. Today probably is in the "supper" computers, making supper-nonsensical "scientific" calculations. One "science" girl on US TV said that space aliens are steeling our discoveries. My opinion is that we are the laughing stock for the space aliens. With this wrong formula the mass of the Sun and stars in binary systems were also calculated. No wonder the density of the Sun is 1.4 tons per cubic meter, when the density of the Earth is 5.5 tons per cubic meter. The real density of the Sun is around 2, 360 tons per cubic meter, so my number is 1,550 times bigger than theirs. Is this a real science or what? Don't forget that at this moment stars are "filled" with hydrogen gas as well, and the stars do not know about it, because of luminaries like Arthur Eddington. Why humans like to put some people on pedestals, and then believe any nonsense from them – beats me. After they put someone on pedestal, he becomes unquestionable, he becomes an authority, and whatever he says should be chiseled on stone. So, the supper great Sir Eddington "reasoned" that the temperature in the center of the stars should be in millions of degree, and the myth of the nuclear reaction where hydrogen is turning to helium in the center of the stars was born. But, the playwright Samuel Becket wrote about one of his heroes~~:~~ ~~"~~: "He reasoned, and he reasoned wrong." To say that stars are filled with hydrogen gas is purely unfounded speculation, and is 100% wrong.

People do like predictions backed by physics and mathematics. If some baseless claim or "discovery" is "backed" by mathematics, it must be true, and that lead to "fake mathematics." The other thing that "legitimizes" some "discovery" is the element of "prediction." So when the timing of the Halley comet's returning was accurately predicted based on Newton's mathematics, it was viewed as a confirmation of the validity of his "gravity." Then the planets Uranus and Neptune was discovered again based on his mathematics, and the reasoning goes that the "influence" of one celestial body to another which is undeniable, should be a confirmation of the validity of Newton's gravity. And that is how a "scientific" nonsense was born. THE GRAVITY ~~AS WE KNOW IT IS~~ OF FALLING OBJECTS TO THE GROUND IS ONLY PART OF THE OVERALL "INFLUENCE" ONE CELESTIAL BODY POSSESS AND DOES NOT EXTEND BEYOND ROCHE'S LIMIT.

Scientists also use in their formulas THE UNIVERSAL GRAVITATIONAL CONSTANT, which by the way is established on the base of the attraction between two bodies made of atoms and molecules. Then in their calculations, they are applying it to "black sphere substance," and then wander what is wrong with their calculations. The "universal" gravitational constant is not universal at all. It is valid only for the ~~second~~ matter #2 which might be around ~~is 2~~ 3% of the mass of the universe.

Almost all that Einstein said is wrong, except for what he receive Nobel Price, which I would term as a lucky guess or he "adopted" (~~it~~ stole) from some other scientists ~~(stole it)~~. Schwarzschild decided to calculate the (density) of a black sphere based on Einstein's fake mathematics. Today his results are accepted as something legitimate, when they are absolutely wrong. Today Astronomical textbooks are filled with ~~90~~ 85% absolute nonsense, and all of them have to be melted back to paper pulp.

SCHWARZSCHILD'S RADIUS OF BLACK SPHERE

Mr. Schwarzschild calculated the radius of a black sphere like this: R = 2GM/c (square). "If the Earth were crushed to become a black hole, it would be about the size of a Ping-Pong ball." BALONEY! Let's examine this "formula." "R" would be the radius of the black sphere; G is a "gravitational constant;" and "c" is the speed of light. "G" and "c" are certified trash; "M" can be only wrong. Although we know that real atoms are "puffed-up" thing, which mean that the actual mass of any atom takes 10,000 times more space (in other books this number is given 100,000). The real numbers for the Earth are 12,756 km diameter against the BS 1,220 km, and that is the ratio of 10 to 1.

The last number "c" is the proverbial speed of light. This is the favorite number Einstein uses to portray himself as some genius, and it really worked for his carrier. When the number 300,000 is multiplied by 300,000 the result is 90,000,000,000. Why Einstein uses this enormous number? The "mathematical crook" uses INTIMIDATION; sudden shock; and "I am smarter than you are" tactics, and it all work~~ed~~s. Believe it or not, the entire world is one big fish which Einstein hooked with his "scientific fantasies" bait.

Einstein, Hawking, Hubble, Doppler, Newton, Eddington, I've had enough of your scientific crap!!! BUT can I save the world from these scientific monsters? It is up to you people; I am doing my part.

BLACK SPHERES CONTENT

~~If we look at our Sun, we know that it has been shining in the last 13.6 billion years. All black spheres in our Galaxy are supposed to behave in a certain way. I do not think that the central black sphere, our Sun, and the Earth are allowed any independent life; any deviations from the "main scenario." All is commanded from the "script" of the Grand Universal Cycle. The fate of all the participants in a galaxy is predetermined and any variations from it are wishful thinking. For~~

Most definitely black spheres are filled with BBPA, but they are not stacked like potatoes. There is a definite structure in their arrangement. When gas enters in a BS, it is dismantled to BBPA. In these cases, we see that from the both polls are ejected gamma rays. This fact can give us some clues about the internal structure.

After the BIG UNIVERSAL AND LOCAL EXPLOSION (BULE) from the ONE OF THE BIGGEST BLACK SPHERE (OBBS), the flying pieces of BLACK SPHRERE SUBSTANCE (BSS) begin immediately to function as BS or maybe not, if they were compacted. So, whether they are ready-to-function BSs or "compacted" (matter # 3); the pecking order is established immediately, because what counts is the INFLUENCE OF THE AMOUNT OF BBPA. The GRAND UNIVERSAL CYCLE (GUC) beguins for this particular cluster of galaxies. Now, how many OBBSs were exploding for the creation of one super cluster is to be determined?

After all the BSs in all galaxies become "functional," they were in the state of the densest state of matter # 2, and later with the constant release of BBPA; their density slowly diminishes. If after the biggest explosion we have only neutron stars and white dwarfs; they have to "recover" and become "normal" functioning BSs. If this is the case, then the first galaxies to appear are not spiral galaxies. But if we accept the hypothesis that the galaxies go through "evolutions" and at the beginning they are lentils (supposition) then become spiral and finally become elliptical. At the end of the cycle all the "matter # 2" consisting of atoms has to be "stashed" back in the BSs through braking the atoms back to BBPA.

~~where, I imagine, the final "purification" of the BSS is done before the creation of the BBS. How many BBSs are created, maybe as many as there are "local groups." Then the proverbial cycle begins again, and is absolutely the same.~~

FORMATION OF OUR GALAXY AND THE SOLAR SYSTEM

There are two versions possible. The first version is that after the biggest explosion the BSs were ready to function immediately; which means that they were not "compacted' from the explosion, but I doubt that. Most likely they were "compacted;" thus not operational and needed recovery time. That gives an answer to the question why our Galaxy began its life 13.8 billion years ago, but the Solar System began its life 4.6 billion years ago. <u>Plus, if the planets and the moons were ready made at this time, and we know that this process of self-creation takes billion of years, then no wonder that from the creation of the Milky Way 13.8 billion years ago to the time (4.6 billion years) the Earth was fully developed 9 billion years have passed</u> ~~The BSs needed "recovery time~~. We could say that the Solar System (SS) began its existence simultaneously with the creation of the Milky Way, but was "dysfunctional." The Sun was back then ~~was~~<u>being</u> a "white dwarf," and the planets were about 1,000 times smaller than now. 4.6 billion years ago the Sun <u>become</u>~~assumed~~ the size that is right now, <u>and</u> let us not forget that BSs once formed and functional remain the same size to the rest of their lives. Same is valid for rocky planets BSs, but the actual size of the planets grows as long as their BSs function. The Moon no longer can grow its BS is dead, but the Earth with functional BS inside does grow~~, but like all BSs the Earth's BS does not change in size~~. The density of the Sun and our BS diminishes gradually with time. The density of the Moon's and Mar<u>'</u>s ~~density~~ equals the density of rocky material plus iron – they are dead. ~~This is the real evolution; not the scientists' illusions. The Sun has been pumping out BBPA as a Solar wind for more than 4.6 billion years; how much is left?~~

~~Let us go to version number one; assuming that immediately after the Big Universal and Local Explosion all BSs which were stars and future planets at this time began to release BBPA. So it was one big mess that is seen in the "lentils" galaxies.~~ Although the Sun was "administering" its "order<u>s</u>;" things were chaotic. As far as Newton's gravity, it did not exist then, and do not exist now, but when people would put it in the trash is difficult to say, as this nonsense has been with us for more than 300 years. I do not know what Darwin would have said about that?

What I am certain is that in addition the biggest BS present which is our Sun a minimum of 200 smaller black spheres are necessary for the formation of all 9 planets, 140 moons and some rocks.

<u>Today</u> ~~T~~<u>t</u>he existence of the "~~second matter~~<u>matter # 2</u>" (atoms and molecules<u>)</u> a<u>re</u>~~t this time is~~ scheduled for "<u>transfiguration</u>~~annihilation~~." <u>This process would take a lot of time. The definite end could be expected when Andromeda collides</u>

128

with the Milky Way. The local group of galaxies at this very moment is in the early stages of congregation, and all of them getting together in one oversized elliptical galaxy, which in its part would get together with other oversize "elliptical" in preparation to become quasar and eventually one of "the biggest black spheres in the universe".

— ELIPTICAL ORBITS

Let's say the following experiment is conducted on a flat dry-lake surface in the desert. To a truck a platform is attached. To the center of the platform a poll is erected. On top of the pall is attached with elastic tether a flying self-propelled gadget. Let us say that we start the rotation of this "mini airplane" which would go in circles around the pall, and then the truck together with the platform would start moving forward. My contention is that as the speed of the truck increases; it would create elliptical. When the truck is not moving the orbital rotation of the flying gadget should be perfect circle. Let us say that this truck is the Sun (220 km/s); the role of the elastic tether is its "influence" or "dictate" imposed on Mercury to orbit around him, and the result is this elliptical orbit. The precessions are result of the "repulsive" zone Mercury has found itself, and moving at the highest speed compared to other planets in the SS.

GALAXIES EVOLUTION

Yes, there is galaxies evolution, but is not what the scientists are assuming it is. They are looking for some kind of evolution after the Big Bang. The problem with this is that this Big Bang is a fiction. Yes, there was a big explosion that created the local cluster of galaxies. The universe has existed forever and will exist forever - it is just going through one and the same never-ending cycle or multiple and identical cycles, which I call the Grand Universal Cycle. It lasts some 20 billion years, and the galaxies are going through ever-repeating one and the same evolution. Should we assume that the so-called "spiral" galaxies are created immediately after the big universal explosion, I doubt that. Reportedly they are only 30% in the universe; what is certain that they become enlarged elliptical; where many spirals congregate and become part of one elliptical. We could say that after the Big Universal and Local Explosions, 30% of the time we have UNKNOWN KIND of galaxies; then the same galaxies become "spiral," and finally the spiral become elliptical. At this moment we are in the phase of initial spiral congregation. Do not confuse the "intruder" with this schedule. If the intruder gets any closer to us it can kill us – through killing the Earth. Keep your fingers crossed.

OBSERVATIONAL CAUTION

I want to point out some pitfalls in "observing" the "universe." At this moment they observe a universe that was created by some Big Bang. In the same time their universe is expanding, or might start contracting. In their minds entire universe was created 13.8 billion years ago, and eventually the stars will burn all their fuel and all this shining galaxies and stars would die. All this is complete baloney! Now we know that all this is false, but there is more to that. OUR GALAXY IS 100,000 LIGHT YEARS across, so anything that has happen before that in our galaxy is forever lost and irretrievable, and that is 13,700,000,000 years of unrecorded data never to be examined! When scientists are looking at remote galaxies they assume that they are looking back toward the beginning of time the "Big Bang." There are two errors in this assumption. First one is that the Big Bang has never happened, and the second is that the distances to these remote galaxies are measured through using two "scientific" nonsense: The first one is the "Doppler Shift" and the second one is "Hubble's constant of expansion". Yes, dear scientists, you are looking back in time, but it is not toward some imaginary beginning, and when trying to ascertain the distance to some remote galaxies do not use Doppler and Hubble's nonsense. I might develop for you more accurate measurement which would be based on the red shift the light acquires going through gas, dust, and particles, but could not be accurate as well – the gas and dust are not uniform medium.

When BBPA are released, the CONNECTING parts are among them, and the formation of absolutely new atoms begins instantaneously and spontaneously. What kind of chemical elements are created depends on the speed of BBPA release, and as direct result of that, what temperature is achieved. We know that in the process of self-creation FREE ENERGY is released. Currently the scientists believe that when looking at a star the elements they observe come from INSIDE the star. Neither the temperature nor the chemical elements are coming from inside the star – they are created OUTSIDE of the star, and the height of the temperature determines what kind of elements could be around. Allegedly when the temperature reaches 11.7 million degrees Celsius, hydrogen explodes in a nuclear explosion, and becomes helium. This might be the reason for supernovas and variable stars.

WHY IN THE UNIVERSE THE MOST ABUNDANT GAS IS HYDROGEN, because it is the easiest self-made element. Second most abundant is helium, and it is not only the second easiest self-made, but is the final result of so many thermonuclear explosions.

It is only logical that the released BBPA should create matter inside of the Earth as we speak. But we can ask another hypothetical question: If certain planet or moon is bombarded from outside with small chunks of "matter number two," could this additional matter makes this planet or moon "bigger" that the

BS inside requires and "dictates?" The answer is no! The newly-arrived matter is subjected to the internal temperature of this particular BS. ~~amount of rocky material is determined by the temperature, and the temperature is determined by the size of the BS inside. The planets and the moons have been hit many times by asteroids. Sometimes these asteroids enter into the mantle, but they cannot create substantial additional mass because are subjected to the "dictating" heat of the BS inside.~~

DISTINCTION BETWEEN PLANETS AND MOONS ON ONE SIDE AND BROWN DWARFS ON THE OTHERS

I have stated before that there is no difference between planets and moons. So when references are made about planets, they automatically apply to all moons, except Io. From the smallest BS to the biggest (1,220 km diameter) which is the Earth's, we have accumulation of "matter number two" proportional to the size of the BS inside, and these are the PLANETS. This rule changes with the brown dwarfs (BD) where with the bigger size BSs begins the process of REJECTING the "matter number two" accumulation, and in some small BD like Jupiter we have interesting "development" in "size" where two equal in size "Jupiter's" would have substantially different powers. This phenomenon is possible only around the size of Jupiter's BS. As the Earth is the biggest possible planet in the universe, we can say that any BS bigger than 1,220 km diameter would create brown dwarf. Happens that our neighbor Venus is that planet / brown dwarf. Officially no measurement of its BS is given, but I can assure you that it is bigger than 1,220 km diameter. Brown dwarfs could never; under any circumstances have a living atmosphere! No matter what is their distance from the star they are orbiting. The scientists are referring about some "planets" as big as 10 times the Earth, but no such planet is ever possible! What they are referring to are brown dwarfs. No spacecraft can land on them, and no astronauts can walk on them regardless of their equipment. All brown dwarfs we have in the Solar System are small, because the next on line after the brown dwarfs are the stars. That means that brown dwarfs closer to their biggest sizes are as powerful as stars – only on a smaller scale.

In comparison between our moon and the Earth, the moon's elements are created at lower temperature, therefore some chemical elements in the Moon should be less temperature resistant than these found in Earth. The chemical composition of Earth and Venus rocky material should be slightly different as well. The higher internal temperature of Venus should preclude the existence of some chemical compounds found on Earth. We are talking about temperature inside the mantle. Surface temperatures are another matter. My suspicion is that the reported surface temperature on Venus of 480 degrees C is ABSOLUTELY NOT from "green-house effect" as the scientists suggest, but from HOTTER MANTLE which precludes the creation of ~~hotter mantle close to the~~ surface crust. This means that the created water is evaporated instantly. ~~of the planet, and because of that, not allowing creation of a crust like the one on top of the Earth. Because of this high internal temperature, ground and underground water are unthinkable.~~ So in a sense Venus ~~do~~ cannot have a crust like ours ever, but very thin and very hot. Once again I am emphasizing the importance of our "moderate" size BS, because we can see from Venus that bigger size BS do not allow the existence of surface water, and even if Venus was occupying the Earth's orbit, living atmosphere would ~~not~~ have not developed, because without surface water there cannot be any "living atmosphere."

    I firmly believe in the previous statements, so may I suggest to the scientists that are looking for exoplanets that they CANNOT SEE the BIGGEST POSSIBLE LIVING PLANET in the Universe, which is the Earth; ~~therefore~~therefore, how they can see smaller? All they see are brown dwarfs!!!, ~~first of all they can see only huge brown dwarfs, and what is absolutely impossible to ascertain the size of a small planet's BS.~~ Basically they are pretending that they are doing some sort of "science," but, hey, they have to eat bread as well. Advice to the space travelers: "Avoid landing on Io, your spacecraft might sink into the lava." Here is a sign for Io: "Avoid walking on the surface of this moon, because it is brown dwarf ~~as well~~, and you might ruin your shoes. In case, ~~if~~ you sink into the lava call for help!"

CORRELATION OF THE BLACK SPHERES SIZES AND WHAT THEY WOULD BECOME

    Before we begin the discussion of this correlation, one enormous inaccuracy has to be addressed. Scientists "report" Sun's surface temperature of 5,800 degrees C. This is the temperature of the Photosphere, but the "corona" which spreads ~~million~~millions of kilometers from the Sun and having the temperature of 1 to 2 million degrees C "does not count." How this could pass as any "scientific" reporting!? This is outrageous!!! These inaccuracies should not be tolerated, because they interfere with the scientific conclusions. May I suggest that no star can possibly have the temperatures that they report (between 3,000 or 50,000 degrees C). These are gross and criminal lies, and you expect from them to explain to you what is happening with our Solar System right now? Are you kidding me? Let us get straighten these star temperatures. First, stars are NOT filled with gas, AND DO~~and do~~ have surfaces. The stars surface temperature is INSIGNIFICANT~~insignificant, it might reflect the internal mechanism work,~~ but around the stars in millions of kilometers the temperature is in million degrees due to BBPA release and formation of "matter number two." These supernova explosions are probably thermo-nuclear in nature. So if you have enormous amount of hydrogen in one place, and we know that hydrogen is quite abundant, ~~but~~and for a~~this~~ supernova explosion enormous amounts have to be concentrated in a particular place; and the trigger mechanism is probably through achieving 12 million degrees C. The blast is absolutely horrendous and devastating for any Solar Systems around. If our Earth is hit by such blast all of its mantle would be strewedthrown into space. At least that is what I see in some pictures. What the leading scientists see in theise pictures of complete devastation of many "Solar Systems?" They see GLORIOUS STAR FORMATIONS.!!! The horrendous blast has stripped the black spheres of many planets and brown dwarfs from the melted mantles and metal. If any planet has its matter number two on top of it suddenly removed; its BS, willould shine like a star. And that is what I see in these pictures of "star formations." We have a case of planetstotal devastation mistaken as a star creation.

The biggest BS would command the galaxies. Each galaxy size would be proportionally bigger asto the size of its the central black sphere is. As the Andromeda BS is quite big, the scientists are wondering why there were no "more" stars in this galaxy. The answer is that there is no "Department of Equal Distribution of Stars" in the universe. The size of the galaxy has nothing to do with the amount of stars in it, but definitely reflects the size and the strength of its Central Black Sphere. The smaller BS would command clusters of stars. After that would be the biggest of two or more stars. Then somewhere down the line would be our Sun. Smaller sizes of BSs would become "brown dwarfs." All these categories "Central Black Sphere;" stars, brown dwarfs, planets and moons refer to one and same thing – namely black spheres of different sizes. Their appearance is different, but their core substance is absolutely the same. Our Sun is not that powerful star and tolerates this "photosphere" on top of itself which is quite characteristic for it. As the size of BSs gets smaller they become brown dwarfs. What is the most characteristic about them is the tolerance of predominantly gas with the largest BDs, and metal and rocky material onappears only on the smallestr BD. The larger BDs are like mini stars, but they are always covered with some gas as well.

The dividing line between planets and brown dwarfs are Venus and Io. Around the dividing lines are formed celestial bodies that cannot be clearly classified, such planet/brown dwarf is Venus. Jupiter is called a planet, but if we call the four rocky planets "planets", then Jupiter do not resemble them at all – it is a brown dwarf. Its surface is probably a living hell - absolute inferno with high temperature and radiation, like the surface of an active volcano. There cannot be any surface water or any living creature at all there. Water, hydrogen and oxygen are easily self-produced by the activity of any BS, but Jupiter's hellish conditions would not tolerate them on its surface. Jupiter has a bigger BS than the regular planets, and that means faster release of BBPA. - A chart should be created reflecting what is produced as the speed of BBPA increases, and what is tolerated on the surface of different sizes BSs. Let's see how the today's scientists see Jupiter and Saturn. For these two "planets" their numbers are wrong.Scientists' "fake mathematical"

133

volcanoes are caused only by the activity of a live black sphere inside of the celestial body, and if the BS is dead, there cannot be any volcanoes what-so-ever. To talk about "gravitational" volcanoes is beyond ridiculous. We are smarter than monkeys after all! Immediately have to be mentioned that only on Earth in the Solar System there are two kinds of volcanoes - universal and subduction.

We have the Solar wind, as if some "wind" is blowing from inside of the Sun, and it is "blowing" the basic building particles of the atom as they are "forced" out. The Sun being way bigger and powerful than any planet's or moon's BSs creates enormous "wind" of particles. The planets and the moons with their smaller black spheres inside have modest release of BBPA that cannot create such a "wind," and this is the reason of being "tolerant" to accumulation of matter number two on their surfaces which is metals and rocky mantles. Why the metal is first layer and the rocky material is second? Because the metal is subjected to double "pull", first is the regular gravity, and the second is the magnetic attraction. The temperature next to the BS keeps the iron melted, and it melting point is 1,538 Celsius. This is approximate number, but quit revealing. We know that around the Earth's BS there are metals, and iron is of the biggest amount. We know also that the surface temperature of the BS is undetermined but insignificant. The heat present on the very surface of the BS is not coming from inside, but created outside by the Special Nuclear Reaction (SNR), and at this location should be the highest temperature in the mantle. The location of the Earth's SNR should be a spherical, and the similar distance compared to the Sun's corona temperature of 1 to 2 million degrees Celsius. Approximate distance might be around 3,000 t0 3,500 km under the surface of the Earth. If iron was not on top of the BS, and was not rotated by the Sun, when the BS was prohibited from rotation; there would have never been the development of life in the universe, because there would not have been any magnetic field to protect the living atmosphere and surface water. If somewhere in the universe there are microbes or worms – that does not mean that in the same environment would exist developed monkeys as well.

When the BBPA are released inside the mantle, and the free association begins instantaneously creating atoms in this SPECIAL NUCLEAR REACTION high temperature is generated. Molecules are created also, and some become gas, which is building up, and going through the volcanoes pushing along newly created and old lava. Let us not forget that any planet (my definition) that has live BS inside GROWS IN SIZE TO THE END OF ITS BLACK SPHERE LIFE, but the brown dwarfs do not (the 4 gas giants)!.

Volcanoes are very important indicators – if they are erupting there the BS inside is working; if volcanoes stop erupting the BS inside is dead, and all plants, animals, atmosphere and water would disappear. There is no difference between planet and moon's volcanoes, universal volcanoes are the same throughout the entire universe, but gases emitted from them are somewhat different for different sizes of black spheres. Eventually a chart could be made that would reflect what kind of chemical compounds come out of what sizes of a BSs. Charts should be made for stars which have the difference of the predominance of specific elements. They are determined solely on the speed of release of BBPA, which is directly proportional to the size of the star and the temperature associated with it – the bigger the star; the faster release of BBPA; and higher temperature. On this issue about measuring the temperature of the stars, scientists are mistaken. Here is an example: The "surface" temperature of the Sun is reported as 5,800 degrees C. I am not interested in this temperature - the real temperature of our Sun is between 1 and 2 million degrees C. First of all, the leading scientists do not understand how the Sun functions and creates its temperature. They have assumed that the Sun is full of hydrogen and some mysterious nuclear reaction is going on in the center. Hydrogen is a gas that explodes at high temperatures, and supposedly stars are filled with it, yet the whole star does not explode!? This is like making explosions inside a warehouse full with explosives. Yes, we are smarter than monkeys, but with how much?

Talking about planets: Two black spheres of equal size should produce the same minerals and mineral compounds. It had been stated that the rocks from our moon are the same as these found on Earth, but I would disagree – they should be slightly different. The temperature inside the Earth should be higher than inside the Moon when its BS was working. Elements and compounds that are sensitive to higher temperature and find themselves in evaporation state should be missing from the Moon's rocks. We know how big our BS is (1,220 km) and the diameter of the planet 12,576 km this is close to the ratio of 1 to 10. Although the Moon's BS is dead – its size probably was around 347 km because its diameter is 3,476 km. These ratios are approximate, and are valid only for black spheres smaller than 1,220 km. Bigger sizes of black spheres produce brown dwarfs and their ratios of BS to overall size are different. The first transitional planet/brown dwarf is Venus. Scientists have never reported the size of its "core," but I am certain that is bigger than the Earth's. My definition for a planet or moon is that a spacecraft can land on it, and astronaut can walk on it. If there are any differences between black spheres – it is in their size, and proportional to that is the speed of releasing BBPA. This is the only condition that regulates what kind of star, planet, or brown dwarf would be created. So we can make new categorization for stars, brown dwarfs, moons and planets only this time around it would be on a REAL SCIETIFIC FACTS.

Let us get back to the creation of metals and rocky material in planets and moons. As I said they should have some differences. These differences should be based on tolerance to heat, and we know that heat is directly related to the speed of the BBPA release. The faster the BBPA are "expelled," more gas is generated because more atoms and chemical elements are evaporated which translates to more volcanic activities. As the temperature increases with the size of the BS, the chemical "stew" (rocky mantle) changes its content as well – only the high-temperature resistant chemical elements remain. Finally, the iron is evaporated as well, and this has happened from the size of Jupiter's BS of 10 Earths to the size of Saturn's BS reportedly between 10 and 20 Earths. These are approximate numbers, because the necessary data is not available – the size of their BSs.

Only on the Earth there are two types of volcanoes, "universal" and "subduction." The universal volcanoes are the same throughout the universe, but the subduction volcanoes could exist only if there is "plate tectonics" in combination with living atmosphere. The gases coming from the are organic compounds that are in the "sub-ducted" plates, and when pushed into the hot mantle the organic compounds create gases that have only one way out – going up to the surface. We extract oil and gas from the Earth's crust, what happens to this oil and gas when pushed into the hot magma? These gases constantly appear in the same place, and volcanoes appear on the site to accommodate the burned oil and gas. Such volcano is probably Fuji in Japan. Firstly the living atmosphere and the Sunlight create the vegetation, and then the internal heat in our planet has created the oil, gas, and coal through heating up dead vegetation in the crust. In this case we could and we should find coal, gas, and oil on Mars, but not where the 4 biggest volcanoes are located, because when they were created – the BS was dead. Let's say that Venus should have lava moving below the surface, in this case the process of subduction should occur, but definitely cannot produce volcanoes because there are no organic materials in the subduction zone to create gas. What comes out of the subduction volcanoes is not that interesting for our inquiry right now. It has to be noted that these subduction volcanoes on Earth are quite a few, and that creates the false impression that the Earth has too many volcanoes compared let's say some other celestial body that has only universal volcanoes. The overall count of volcanoes on the Earth compared to another celestial body with only universal volcanoes, might lead to some erroneous conclusions.

There is third kind of volcanoes and they are the shield volcanoes. These volcanoes are universal, but without the gas. As the BS inside our planet works, it creates new matter all the time, and this newly created matter comes out on the surface of the planet through this and other volcanoes. Such place is the Atlantic Ridge where newly created matter is pushed to the surface of the planet, but some of this surface happens to be under water. As new matter is created and pushed to the surface of the planet, it INCREASES THE SIZE OF THE PLANET!

Hawaiian volcanoes are universal volcanoes. I would say that it is not many volcanoes, but only one. On the Big Island of Hawaii there are ostensibly many volcanoes, but they are just many vents of only one volcano. The Earth crust is moving, but this volcano's vent remains in the same spot, and that creates the impression that this volcano is moving, when actually the Earth crust is moving. This is a well known fact, but what we have to examine is this "tubing" or "funnel" that was created "permanently" from deep inside the mantle and serves as conduit for newly created matter and gas to come to the surface. I imagine that this is some sort of funnel turned upside down that "captures" newly created gases, which on its turn push upwards melted matter, as the newly created matter creates pressure inside the Earth, and the material is being "expelled" out through the volcano. Only t~~T~~he universal volcanoes create the impression that they are "moving" through the crust. ~~This phenomenon is valid for all universal volcanoes.~~

The gases coming from universal volcanoes are very revealing as to what elements are self-created inside the Earth. Scientists are saying that some elements like gold have been created through supernova explosions, thus have come to Earth from outside – I strongly disagree. Almost all elements are created right here inside the mantle, because almost all matter on top of this ball of 1,220 km diameter is locally produced by the Special Nuclear Reaction (SNR). On TV I was watching a Japanese man boiling eggs in water running next to his house. Water that is warmed by the Earth's nuclear reactor! Hear is the question: What is the temperature inside the mantle where SNR takes place? ~~Are the first 30 elements from the periodic table created in temperature between 5,000 and 6,000 degree Celsius, and is this temperature accurate? Sulfur becomes something like gas, and in that condition comes to the surface of the Earth.~~ I have smelled sulfur in one place where there was no volcano. That should mean that the Earth's crust at this place is very thin.

~~If we look at the organic compounds in the Earth's crust, I think that they add insignificantly to the size of the planet, as they could exist only in the crust.~~

I have the suspicion that the "sunlight" adds to the size of the Earth, but so far there is no evidence. I would point out that the Solar Wind is particles, and the light that comes to the surface of the planet should be particles. My theory is that matter and energy are inseparable; ~~therefore~~therefore, there is no energy without mass. Only on the Earth ~~T~~this "new" mass is created by the living atmosphere and the light from the Sun, and does not exist in any other planet or moon in the Solar System except on Mars, because it used to have a ~~is created by the~~ living atmosphere. These newly created chemical substances have in them the combination matter plus energy from the Sun, and that is how we derive energy from burning wood, coal, petrol, and gas.

Sulfur dioxide is prevalent on Io, and exists here in Earth's volcanoes, but as I have identify Io as a brown dwarf – there should be some difference of gas content in ~~from~~ the volcanoes of these two bodies. A bran new investigation has to be launched in this direction until its full explanation is achieved. This investigation should be part of answering the question what are the self- produced elements for different sizes of BSs. ~~As the new chemical elements are self-created, new chemical compounds are created from them.~~

BLACK SPHERES AND ELECTROMAGNETIC FIELDS

If one planet is to have an electromagnetic field, three requirements have to be met. The first one is to have "live," therefore functional black sphere inside. Second one is that this planet has to rotate (Mercury and Venus are prohibited from rotation around their axis by the Sun), and the third requirement is that its BS should be at its rightful place – the center of the planet (Mars' BS went from the center to being close to the surface 700 million years ago; our BS is off of the center of the Earth and as a direct result of that our magnetic field is not functioning properly. How exactly magnetic fields are generated? Black spheres are magnets with a polarity – north and south poles. We also know that the Sun is making "matter number two" to rotate around the black spheres of the planets and brown dwarfs in the Solar System except for first two planets which are blocked from rotation. The iron is subjected to double gravity – regular and magnetic, and that is the reason why is first layer on top of the BS. That is how a naturally made electromagnetic motor is created. The iron around the BS is not evaporated because the temperature there is relatively cooler, as the highest temperature is somewhere in the middle of the mantle.

The "matter" planets are made from (the mantle) is not accrued from outside, but is self-created from the basic building particles of the atom released by their particular BS. Their creation might have begun approximately 13 billion years ago, or later, but not 4.6 billion years ago, because the planets and the moons were already developed at that time. The Earth developed its magnetic field slowly with the creation and accumulation of more iron around its ~~the~~ BS.

The magnetic field is like a cocoon that protects the living atmosphere from the Solar wind. Without it there won't be any life on the Earth, and there won't be any surface water. If something happens to this field which I call ~~it~~ a "shield," all water, air, and life will disappear, and is disappearing right now. Don't be scared, at this very moment, our "beloved" black sphere is off of its rightful place which is the center of the Earth, and we know that this is one of the 3 absolute requirements for el-mag field. With the central core of Earth's electromagnetic motor off of its central location some 400 km (size of the core 1,220 km). Supposedly we are going through so-called "pole-reversal." Forget the pole reversal, we have in our hands "central core displacement," and that is scary! This "reversal" might or might not happen, but this "transitional" period of uncertainty is unavoidable and unnerving. I would call it: "Pole-tilting and leaving it in 'whatever' position." We were wondering what caused Uranus' rings tilted perpendicular to the orbital rotation, or all these angles between rotational axis and magnetic orientation. I have already reported my discovery that the Sun "makes" all planets rotate on their axes, so we can continue to rely on this service, but the poles would remain as this "intruder" leaves them when its unwelcome influence would end. Should the angle between axis if rotation and polls change? Depends only on the "intruder" - how he would leave the "scene" of the crime after leaving the Solar System. We are losing precious atmosphere at this moment and this can last some time. We could know how long would last only after locating visually "intruder," and see its speed and trajectory. At this moment, as other scientists are not trained by me; only I could interpret the data properly.

Let us elaborate on this one. The Sun "orders" the planets to orbit around it, and it does not "pull" on their black spheres, so we have here "intruder" who cannot affect substantially the Sun, but sure enough can displace the planet's BSs. What is peculiar in this tug of war between the Sun and the intruder is that regular mass of the planets which is the mantle gets one treatment, when the black spheres which are 700 times more denser – different. Is the different treatment being a result of different density, or the fact that the BSs are magnets, but the regular matter is not? The Earth's magnetic poles are slowly moving. These movements are caused by the intruder, and we can pin-point to the direction from where this influence comes from. Let us remind ourselves that this "influence" might not be from an intruder, but from a star outside of the SS.

The Sun blocks the rotation of the BSs inside the planets and "orders" the rotation of the melted metal and rock material for all planets, except for Mercury and Venus which rotations around their axis is prohibited. Deprived from rotation, they cannot have magnetic fields. Mars does not have one because does not have live BS inside, but continues to rotate because this is how the Sun dictates, and lost its atmosphere as a result of losing its magnetic field. Today's scientist will tell you that gravity holds the living atmosphere, and that is NOT TRUE.

The Sun is who dictates and enforces the "no-rotation order" for the moons. This and other intricate details have to be understood, so we would have better understanding of the inner workings of the Solar System before the intruder's coming in here. If our Moon was allowed to rotate on its axis, it could have "moonlets," but they could have gotten so close to the Earth that the Earth's influence/gravity might have caught them. Sounds like an intelligent design isn't it? Could this "intelligent design" happen by itself, by the system, or by God?

HOW TODAY'S SCIENTISTS "PRETEND TO STUDY" THE STARS?

Their first mistake is the assumption that the stars are made of atoms – they are not. Their second mistake is the assumption that stars are filled with hydrogen gas, and some nuclear reaction is going on for billions of years in their centers where hydrogen is turns to helium. Then they look at the stars and see different colors and strength of light – continuing ancient Greek tradition, but all these are dead-end observations that cannot lead to any real discoveries. In the same time, they are making laboratory discoveries about spectral lines of different chemical elements, when the stars do not posses any chemical elements, but are surrounded by them. All this is made to look like some sort of scientific progress, but it is not. Next they are attaching different gadgets to their telescopes (spectrographs) and recording different chemical elements around the stars; in the same time they measure particular star's surface temperature through looking at their surface color – all this is fruitless "games" in which they pretend that is some sort of science.

They have elaborate diagrams – Herzsprung-Russell diagram; Henry Draper catalogue of spectral sequence. All this is designed to intimidate you, and to look highly sophisticated, top of the line, cutting edge scientific - nonsense. I can go in detail with any of their pretences, formulas, graphs, and plenty of fake physics and mathematics. One has to be pathetic liar to be in this field, and have to accept all the previous lies, misconception, deceptive physics and cooked-up mathematics. The stage is set and if one wants to be in this field just have to continue the "production" of lies. It's like sausage making tradition. One should learn all the previous nonsense and add to it, and that is what Stephen Hawking did. His contribution to the already large body of "scientific baloney (knowledge)" was that the black holes emit radiation – therefore one day they will disappear like "melting away" – depressing and disgusting stupidity. Instead of getting smarter, the human race is becoming smarter with a minus sign in front it, but that is not what the God's plan is. God loves you, want you to succeed, and forced me to discover the universe and GIVE it you – be brave and forge ahead! But on a personal level, we are still too close to our predecessors – advanced monkeys.

Let's go back to the astronomers, they know only one "matter," this is the matter the ancients new, and we live in it. So their investigations can lead them only to wrong conclusions!

The initial model of inquiry is very important – today's Astronomy scientists are operating with multiple wrong models. All is "legitimized" with layers upon layers of fake assumptions. The newcomers in the field imitate the previous nonsense-talkers, and the tradition continues. The leading institutions in this field are staffed with true believers, non-thinkers, and people with connections to the security agencies. So who decides who should be a scientist? Can the real science "function" this way? The scientists become something like "two for a dollar", or something like disposable dippers. Here are the credentials of S. Hawkings: in 1962 he earned his doctorate degree at Cambridge; in 1974 California Institute of Technology; fellow at Royal Society since 1974; he received Eddington Medal of the Royal Astronomical Society; Dannie Heinemann Prize of the American Institute of Physics; the American Physical Society; Maxwell Medal and Prize of the Institute of Physics; Einstein's Medal: in 1979 he was elected to Newton's old chair Lucasian Professor at Cambridge. With all these impressive credentials he was one of the scientific mafia boys – licensed to continue the tradition of spreading scientific lies throughout the world. He also "discovered" that after the Big Bang there were some mini black holes – smaller than basic building particles of the atom. The moron thinks that we eat the tomatoes with the leaves! By now you might have gotten some sense of what the "scientific mafia" is.

EARLY DISAPPEARANCE OF BLACK SPHERES

We have two cases of early "disappearances" of black spheres. One is our Moon, and the other one is Mars. The disappearance of Mars' black sphere could be directly traced to the last volcano on this celestial body – Olympus Mons, but the time of disappearance of our Moon's black sphere remains a mystery. My investigation into the nature of the black spheres couldn't be thorough for many reasons, but I should not be blamed for "incomplete" report, due to the enormity of the task, and the circumstances I have to deal with.

Why our BS is still working, when some others are dead? Notice that these dead planets and moons are of smaller size, therefore their black spheres simply spent all of their BBPA. Here is some drama about the Mars' black sphere disappearance. It "died" quite recently, 30 million years ago, but the atmosphere, surface water, and the eventual life on it have disappeared 700 million years ago. It was a high drama. For example, from 700 million years ago to 30 million years ago, Mars was wobbling ~~real~~really badly around its axis. Let us not forget that the BS was 1/10 of the size of the planet, but its density was ~~maybe~~ 700 times bigger. The black sphere was way out of its ~~proper~~rightful place which is in the center of the planet, instead the BS stay right under the surface of the planet starting 700 million years ago and staying there 670 million years until its final days 30 million years ago. What exactly happened? Some intruder like we have right now might have to be blamed for "displacing" Mars' black sphere off of its place. This brings another question: Why black spheres move so easy from the spot they have to stay? First of all, they create heat in thousands of degrees around them, so they are surrounded by melted metal and rock, thus easy to move through this melted substance. Second they are hundreds of times denser than surrounding matter – the current number for the Earth's BS is 700 times denser than the stony mantle, and third they are magnets. If we try to explain through MAGNETISM, the displacement of BSs; then the reasoning should be that the BS magnetism might be 700 times bigger than this of the mantle. In this case, if huge magnetic force is applied to both of them, most definitely the matter # 1 in the BS would act more rigorously, than the matter # 2, which is the mantle. Also, may be the grip of the Sun on the BS might be way stronger than the mantle, and when the tug of war between the Sun's influence and the intruder's happen, a planet might be "dismembered" (if we could use such term). Fact is that Mars' BS was forcefully moved from the center of the planet all the way under the crust.

WHAT HAPPENED TO MARS COULD HAPPEN TO THE EARTH

What is happening right now with the Earth displacement of its BS is similar to what happened on Mars, but in a smaller scale. Mars' BS was displaced all the way under the surface, when the Earth's BS at this time is some 400 to 500 km off its base, but if this monster gets any closer to us - the displacement might be catastrophic. We have to elaborate on this "catastrophic" displacement. The question is: As the BS is removed from the center, would it return back to where it belongs after the intruder leaves the Solar System, and is there "distance of displacement to a "point of no-return?" Let us hope that the current displacement of 400 to 500 km does not increase significantly. If this distance increases, the Earth's wobbling would increase proportionally; the magnetic field would get worse; the loss of atmosphere would faster; the weather would worsen; and the solar radiation would increase. Let us assume at this moment that the "intruder" causing us all this grief might be another 200 or more years around. This is an emergency! We must locate the intruder as soon as possible, and have definite answers to this life- and- death questions!

Why there are 4 enormous volcanoes in one place on the surface of Mars. Let's first look at normal BS activity which are on the Earth, and compare it with the abnormally big volcanoes on Mars. When a BS is in the center of a celestial body, its volcanoes are dispersed "evenly" throughout its surface, and their activity are pretty much "even." This is the case with the Earth, but things were different on Mars. We have 4 volcanoes that are the tallest in the Solar System. When Mars' black sphere was yanked out of the

143

center of the planet, it went under the so-called Tharsis Bulge area. Obviously the Mars' BS stayed in this area under the surface of the planet from 700,000,000 years ago to as recently as 30,000,000 million years ago. The BS did not return back to the center because it has crossed the point of "no-return."

Mars' BS was very close to the surface of the planet and built these enormous volcanoes one by one. The first volcano was Arsia Mons, which is 9 km tall. At this time Mars atmosphere was in the process of disappearance because the magnetic field was dislocated. Looks like Mars used to have a lot of water, and at this time the surface water was frozen. So Valles Marineris was a vast frozen water expanse. It is easy to see that this "canyon" begins from this volcano. The enormous amount of hot lava created an enormous lake of hot water, which I assume carved the "canyon" because of the rotation of the planet. As I said the surface of Mars was covered with ice, but as the black sphere found itself under the surface, and pumping all the hot lava and gases – creating the first huge volcano. The ground around it was very hot, because of the functioning BS very close to the surface, so was created permanently hot spot which was a constant source of hot water that created Valles Marineris going through the deep ice. As the hot water was forging the canyon, we can see several places where it pool up temporarily. The hot water carved this canyon. Why the water froze on the surface of Mars? After the BS was substantially moved away from the center of the planet, the magnetic field no longer covered the planet properly or disappeared instantly, and let us remind ourselves what today's scientists cannot comprehend that once the magnetic field is missing – the disappearance of atmosphere follows immediately, and for whatever reasons the water freezes. At this time there was not FUNCTIONAL atmosphere; whatever air there was no this planet leaked out first, so water remain last to disappear slowly into interplanetary space but was frozen. Scientists have this Steven Bolzman's law of how air molecules have to have some speed in order to get out of the gravitation, and the living atmosphere keeps that temperature mild, but things do not work that way in the real world. The Solar wind particles have enough energy to nock air molecules out of the planet's gravitation. The scientists believe that what holds our atmosphere is the gravity. "The Moon cannot have atmosphere, because do not have strong enough gravity." My answer to this is that the Moon does not have atmosphere because does not have electromagnetic field. Here are the reasons why the Moon cannot have natural atmosphere: First one is that its BS is dead, and even if it was alive, all moons are not allowed by their planets to rotate, and without rotation there cannot be electromagnetic field. I am asking the scientists: If Mars used to have living atmosphere, why it disappeared? My answer is, because it lost its electromagnetic field, which is the guarantor of the existence of living atmosphere. Mars still have gravity, but the atmosphere is gone. They look at Venus and call it "atmosphere". Firstly, this is not a living atmosphere, and it has not disappeared because ~~of~~is constantly suppl~~ied~~y ~~of~~with gases from the interior. Second they are victims of their semantic inaccuracies – calling apples oranges – they use the same word "atmosphere" for Venus' gases, Mars' lack of atmosphere (calling it 1% of ours), Sun's outer layers are "atmosphere" – I strongly disagree with this inaccurate terminology which leads to illogical conclusions.

Any way, let's go back to our 4 volcanoes. These volcanoes are so huge for ~~one~~two reasons: The first one is, ~~and that is~~ that the displaced BS was right under them and the gas and new matter generated from the self-creation of different atoms and molecules was vented mainly through one of these volcanoes at the time; the second reason is that, ~~and~~ there w~~as~~ere no atmosphere~~ic~~ to erod~~ing processes to lower~~ their heights. We can see the path the BS has taken staying ~~from~~ under one at the time of these volcanoes. 700,000,000 years ago it went under the first of the four volcanoes - Arsia Mons 9 km tall which is 30 degrees south of the equator. Mars' BS stayed under this volcano for 400,000,000 years. Then went under the second volcano on the very equator - Pavonis Mons 7 km tall, 300 million years ago, and stayed there 200,000,000 years. Continuing somewhat in straight line the BS went 30 degrees north of the equator under the volcano Ascreaus Mons 18 km tall 100 million years ago, and finally the last "appearance" and "performance" Mars' BS made was under the tallest volcano in the Solar System Olympus Mons 21 km tall. If the last eruption was 30 million years ago - then at that time Mars' BS "died." The reason traces of a nuclear explosion was reported on Mars is because the BS w~~as~~ere too close to the surface of the planet., and creation of matter # 2 is A SPECIAL NUCLEAR REACTION which released this element on the surface of the planet.

BLACK SPHERES THAT PLAY POSITIVE ROLE IN OUR LIVES

Black spheres are not exactly "friendly" to life, but here it is, we live on a planet which is like paradise. So, obviously, there are circumstances where black spheres create conditions where life can thrive. One ancient Grecian playwright refers to the Earth with the words, "mother of us all." Couldn't agree more, we all have two mothers, one that gives birth and raises us, and then the other "mother" - Mother Earth that shelters, feeds, and gives us everything we need to survive. Enough poetry, let's see what the black spheres are doing for us. There are many BSs that affect our lives, but two of them have the biggest immediate impact on us, and they are, the one in the center of the Earth, and the other one is the Sun. Overall they have been good for us. Crops have been very important to us, and somehow we have survived. Our planet is located in absolute precision from the Sun. Maybe some people are afraid from volcanoes, and rightfully so. Volcanoes are dangerous to be around, and if they are too big, they can create a major problem with relatively long term air pollution. But here is the naked truth, we should celebrate when the volcanoes are erupting, because this is a sign that our "beloved" BS is alive, and will take care of us. The death of our BS spells the death of our planet and everything alive on it – only the space aliens can hop in their UFOs and get some other place – we neither have reliable planetary transportation, nor we know where to go.

Why our BS has to die one day? ,-Bbecause it has a limited supply of basic building particles of the atom, and it has been releasing them in the last (who knows) 10<s>5</s> or more billion years. Let's do not forget, our Moon is dead, Mars is also dead, but our Earth is still alive, and that is good for us right now. Could anyone make some machine that can work nonstop for 4.5 billion years? That is what these two BSs <s>are</s> <s>were</s> doing for us. The Sun has been releasing these BBPA with absolute precision. I have heard reports that the Sun is shrinking, and here is my question<s>this have been one of the questions I have</s>. If this is true, then the Sun in the distant past might has been bigger in size than right now, but at this moment I believe that black spheres are like boxes that DO NOT CHANGE THEIR SIZE, but one never knows when they would be completely empty, and they do not melt like ice as <s>the ignoramus</s> Hawking <s>is</s> claims<s>ing</s>. The report that the Sun is shrinking might be that the Sun might have temporary variations in size, and like a HEART pumps out BBPA, but in long run <s>overall</s> remains the same size. This <s>is only a supposition that</s> has to be confirmed or denied based on observations.

THE SUN AND ALL STARS

The Sun is a star, and all stars are black spheres. As such they release basic building particles of the atom. When released, these particles are in the state of free association where enormous amount of free nuclear energy is released. This nuclear reaction is unknown to the scientists yet. The Sun's surface temperature is insignificant. There is no nuclear reaction going on inside of it, as the scientists believe. All energy of all stars is a nuclear energy which is 100% generated outside. <s>of any star.</s> It has been reported that the Solar wind consists of protons and electrons. As you can see the first creations are protons which are more sophisticated particles of the atom<s>s</s>. Then the most abundant chemical element in the universe is hydrogen which is the easiest to SELF-CREATE<s>be made</s>. The second one is helium with two protons and two neutrons, but this "logic" breaks at next three elements. Why this is happening? This is a home work for the scientists.

Scientists believe that there is star formation in galaxies at all times. I strongly disagree; there cannot be such formation. All stars are created at the beginning of the Grand Universal and Local <s>Cycle</s>Explosion. All stars are pieces of One of the<s>the</s> Biggest <s>Local</s> Black Spheres (OBBS). As the Big Universal and Local Explosion happens; pieces flying from the explosion become galaxies with stars in them. <s>–</s>These stars go through "evolution" in the frame of this Cycle<s>, and eventually are recycled together with the local galaxies and everything in them</s>.

146

Scientists erroneously believe that all started with some "Big Bang" and evaluate every observation from this point of view. They "say" that our Sun is one of the stars that are in some "main sequence," when the stars are not filled initially with hydrogen gas, and no nuclear reaction is going on inside them. I am sorry to say this, but all this is crap.

Let us go back to the stars. Our Sun is a small and therefore "weaker" star, but isn't that what we need? The difference between our Sun's with yellow-red color and other bright and bigger stars is caused by the different speed of pushing out BBPA. The slower rate of pushing them from our Sun allows the creation of Photosphere, where the way faster rate of expulsion in the bigger stars "blows off" the formation of this Photosphere.

At this moment if someone says that the Sun is not filled with gas might be labeled an idiot. Let's see how this "hydrogen gas" got into the Sun? Sir Arthur Eddington "studied" the "internal structure" of the stars; according to UNIVERSE 2020 by Smithsonian. Why don't they speakay the truth, he made all this up. How can he study something that nobody knows anything about? "He guessed it, and he guessed it wrong." The play writer Samuel Becket had the phrase: ": "He reasoned, and he reasoned wrong." I do not think that Mr. Eddington even has the ability to reason, but he teaches the world through his guessing, and no wonder, he was friend with the biggest guesser ever in the universe Mr. Einstein.

Here is what the contemporary scientists know about the Sun, which I would characterize as next to nothing. Smithsonian Universe 2020 p104: ": "The Sun is a 4.6-billion-year-old main-sequence star." The Sun is definitely as old as the Milky Way, and that cannot be 4.6 billion years, but more like 13.68 billion years. The quote continues: "It is a huge sphere of exceedingly hot plasma (ionized gas) containing 750 times the mass of all the solar system's planets put together." The Sun is a black sphere and there is no hot plasma inside it. Here are some numbers: the Photosphere has the temperature of 5,700 degrees Celsius, but the black spots have the temperature of 3,500 degrees. These black spots are the actual surface of the Sun, and simultaneously the surface of a black sphere. Where this 3,500 degreesdegrees' temperature came from? The Sun is covered with its first layer of "fire" which is 300 to 400 km. from the surface. The high temperature around the Sun is created as the matter number two is created which consists of plasma, atoms, and molecules. Cannot be more emphatic on this one - the energy created in this process is absolutely FREE. There is no conversion of matter to energy ever anywhere in the universe! The surface of the Sun is clearly visible and is black sphere, but the scientists say that the Photosphere is the actual "surface" because in their imagination it is filled with gas. I have made an important discovery: There is no energy without matter associated with it, that means that the light, x rays, gama rays have mass.

From the previous quote, the statement that Sun's mass is 750 times bigger than the mass of planets and moons is inaccurate. I have calculated roughly the density of the Earth's BS. Assuming all BS have approximately the same density right now, then the density of the Sun is 2,365 tons per cubic meter – when the scientists are giving the absolutely ridiculous number of 1.4 tons per cubic meter. Calculating roughly the ratio between the Sun's mass and the mass of all planets and moons; I came up with the number of 100,000 to 1, when their ratio is 750 to 1.

EARTH'S BLACK SPHERE, VOLCANOES, AND GROUND WATER

Because our BS works right now; it has kept the interior of our planet hot, and this is one of the necessary conditions for having ground and underground water. If this heat in the interior disappears, the ground and the underground water would sink into interior. If the interior of our planet was hotter than it is right now, there would not be any crust and water; it would be like Venus with surface temperature of 480 degrees C. Plate tectonics and volcanoes are direct result of the heat coming from the BS activity. The surface temperature of the BS is insignificant – we know that from the Sun. In order to have some sense where would be the highest temperature in the Earth's mantle; we should look at the highest temperature created by the Sun. After all the Sun and Earth's BS are made from the same Black Sphere Substance. The temperature of the photosphere is 5,700 C, but I do not think that it would worm us here on Earth some 150,000,000 kilometers away. Another words if this 5,700 C is the only temperature of the Sun, then we would freeze to death. The true temperature of the Sun is about 1,000,000 to 2,000,000 C in the corona which extends millions of kilometers from the Sun. So, why the scientists report Sun's temperature 5,700 C? Because the scientific mafia like small children have created their favorite toys of some system for classification of the stars which dates back to ancient Greeks, and do not want to give it up. So far all these nonsense have served them well, but now we are in real crises, and the established nonsensical fake science of Astronomy is falling apart – good riddance!

The creation of atoms and molecules in the planets mantles create heat. This "reaction" is nuclear in nature, but does not happen on the surface of the BS but extends into the mantle, and that is how the mantle stays melted. This energy is free, and matter and energy are inseparable. If the energy in the universe was not free, then there would be no universe, and if someone asks what happen to the universe, we have to answer, it ran out of energy. That is what the scientists are expecting to happen after their imaginary Big Bang, and currently they live with the erroneous assumption that matter turns to energy and is forever lost. What supposed to go to the trash is e = mc2, and scientists should find the missing mass from the two hydrogen atoms that become one helium, and not claim that the missing mass has turned to energy. Enough mathematical gymnastics, it is time for the truth to come out.

THROUGH EXAMINING THE GASES COMING OUT OF THE UNIVERSAL VOLCANOES; WE CAN LEARN WHAT ELEMENTS ARE FORMED BY OUR BS. Looks like, all elements from the periodic table are produced right here inside the mantle of the Earth!!! Let us find the answers to these questions: Why sulfur is so prominent, as well as oxygen? How much water is produced? Earth's volcanoes produce sulfur dioxide; so are Io's. The predominant gases coming from the most volcanically active Jupiter's moon Io is sulfur dioxide. A remark has been made about the enormous amount of lava coming from Io's volcanoes. I would caution that this seemingly "big amount of lava" is not new material, but old one which happens to be in the way of the gases coming out.

EARTH'S ELECTROMAGNETIC FIELD IS OUR SHIELD

A warning to the human race: If you listen and believe what the leading scientists are telling you, you might perish because they do not know that much!! Today the Earth's BS is in trouble. Firstly, scientists do not have a clue what is going on. Second, they do not have the knowledge and the equipment to monitor the INFLUENCE OF THE CELESTIAL BODIES. As a result of that, they do not know what is happening to the Earth. I know why the Earth at this moment is wobbling - because the BS is off of its place – the center of the Earth. There is another reason for this wobbling. Maybe the intruder might change the incline of the rotational axis which at this moment is 23 degrees, and might have ~~. This~~ happened through previous "bad" influence. What if the current intruder affected th~~is~~e rotational axis? Keep in mind that this would be very dangerous if done suddenly because ~~of~~ the oceans would spill.

Today the scientists erroneously believe that the Earth and the Moon are in some sort of binary system. All this is coming from Newton's inaccurate physics and mathematics, and his wrong assumptions about the "gravity." Information for the scientists: The whole universe is one SYSTEM in which smaller and smaller systems operate – one inside the other. Nothing happens by chance – all is predetermined. The universe is the all-encompassing system with smaller systems in it.~~-~~ A cluster of galaxies is a smaller system that obeys the bigger system. Each galaxy is a system into the cluster system. Our Solar System is part of our galactic system, and finally the Moon orbiting the Earth is the smallest ~~one more little clog of this enormous and coordinated~~ system within the larger Solar sSystem~~s~~. So when the scientists are telling me that the Moon is moving away from the Earth 3.3 centimeters each year, I do not buy it. The Moon has assigned place and is not allowed to move away from the Earth except if it is in the scripture of the "Grand Universal Design~~Cycle~~." ~~-~~ I want to know how they arrive to this conclusion!? The biggest BS in our Galaxy (Central Black Sphere) ~~which is at the center~~ – " "commands" all the black spheres in the disc of the galaxy. What the scientists see? They look for Newton's no existing "gravity" to hold the ~~between~~ stars inside the Galaxy. What is happening in reality is that the CBS commands the stars like the Sun, and the Sun on its turn commands the planets. Looks like the Sun does not interfere with the planets influence, but this is not completely true. Mercury and Venus are prohibited from rotation around their axes. Planets orbit the Sun, when the moons orbit the Sun through orbiting the planets. So who prohibits the moons from rotation; when if a celestial body is a planet would rotate, but if it is moon would not? My answer is: That is how the "~~System" The~~System" The "Solar Systems" in our Galaxy orbits the CBS where only the star that is in the role of the Sun actually orbits the CBS, but the planets inside do it through orbiting the "Sun." The orders are: ~~This command is very peculiar - it does not go to all black spheres. It goes to the biggest ones and the order is simple:~~ "Orbit the center at this prescribed speed." Now, I am using the word "order" which might be quite inappropriate. This is a "force" of the "influence" which translates to: The larger BSs influence the smaller. Then the ~~Ce~~entral BS does not interfere in the field of influence of this BS, so in our Solar System all smaller BSs orbit the Sun and not directly the CBS. Now we are to examine the Sun's "commanding influence" which resembles the CBS commanding influence, but it is somewhat different. The Sun "orders" all planets to ~~rotate~~orbit around it at preordained ~~orbiting~~ distances and speeds. ~~for the planets.~~ So, Mercury has taken the closest and the fist available orbital spot. It cannot be any closer or farther from the Sun. ~~What the~~ Newton's gravity has nothing to do with i~~ would say about it? t Mooohhaa~~. There is no such thing as "Newton's gravity" – mistakes do happen.

Let's go to the rotation of the Moon orbiting the Earth. The scientists believe that the Earth and the Moon are some sort of a "binary system" which center is somewhere inside the Earth. The reality is that the Moon has to orbit exactly around the center of the Earth's BS, but right now we have DISPLACEMENT of the center, and this might make scientists' erroneous assumption somewhat "correct!" As far as the orbiting around the Sun, only the Earth has been ordered to do so, and the Moon is ordered to orbit the Earth and not directly the Sun, but it is orbiting the Sun through orbiting the Earth. These are the "laws of the black spheres influences" which Newton was not aware, and Einstein together with his protégé Hawking did not have a clue about it.

Isn't it remarkable with what precision the Earth is rotating around its axis? Basically our clocks are set exactly on 24 hours for one rotation. What Einstein would say about it? I hope that the human race understands, once and for all, that this man was not a genius! He "invented" that "time dilation," and now as we have the BS of the Earth off of its center, his idiotic ideas sound like "science." Keep in mind that the Earth's wobbling should be very modest compared to Mars' wobbling 700 million years ago. Although Mars' wobbling was the most severe possible, it did not affect Mars' rotation around its axes consequently. The rotation of Mars did not change because have nothing to do with its BS dead or alive – it is "ordered" and "enforced" by the Sun.

Humans, wake up to the reality that this science has been taken over by a special kind of "scientific mafia" which commands thing from governments, institutions, commercial interests, educational dictatorship, and "uniformity" of some "collective" thinking that is killing the real science.

RECOMMENDATIONS TO THE SCIENTIFIC COMMUNITY

When observing the universe, please record the actual observations first, so that we can have some unbiased cache of information for all who might have some new and different hypothesis or theories from the ruling institutional mafia, otherwise the current continuation of misconceptions will be prolonged. The efficiency and the advancement of science will be greatly improved. I am not saying that the particular observer should not render his or her opinion, but we have to separate facts from opinions. Somehow these fake mathematical formulas have to be exposed – like the requirement for some "N" for Bode's law. Obvious facts are suppressed so that some nonsensical theories are not contradicted – clearly visible surface of the Sun, and disregarding the corona's temperature, because of some "luminosity." ~~Case in point: The claim that the Sun does not have actual surface, when it is easily visible!? One x-ray picture made in 1973 where a big "black spot" in the form of Italy where the EDGE OF THE SURFACE of the Sun is quite visible, but such finding would contradict Sir Arthur Eddington's nonsensical theory that the Sun is filled with hydrogen gas. In the Wickipedia he is called "one of the greatest astronomer ever lived." "Doppler's effect" is another complete nonsense that today's scientists subscribe to. What would it take for these scientists to admit that they are wrong?~~

Eddington, Hawking, and Newton probably are the pride of Great Britain;, I do not know anything about Max Plank, only that he described the first second of the Big Bang. Should the Germans rename the "Max Plank Institute?" How about Edwin Hubble, on his name is the "Hubble Telescope," but there should not be a "Hubble's law" because the universe is not expanding at all! Czechoslovakians are probably proud of Mr. Doppler, but what I am supposed to do?

ERIS

~~There is a very interesting small "planet" called Eris. Supposedly it is a little bigger than Pluto like 2,326 km diameter. As I am examining the negative influence on the Solar System, and the bad influence on our planet, and I am trying to find from which direction this influence is coming; and I am going nut! WHERE IS THIS CELESTIAL BODY THAT CAUSES US TROUBL? I expected the size of this "intruder" to be around 300, 000 km diameter if it was BD in the Oort cloud. But if it is made from "compacted" BS then something the size of Eris could fit the bill, because the white dwarfs are 1,000 times stronger. I am not saying that the culprit is Eris; only to point out that size and power are different, and from something small some enormous influence could be emitted.~~

SPACE ALIENS AND HUMANS

We live OUR lives that are from the script of the space aliens's scenario – we. ~~We~~ do not write our history through the actions of our free will, but "perform". ~~T~~the "history" which~~ go through~~ ha~~s~~ve been written ~~for us~~ upfront for us. We are only the puppets in their show~~a show~~, and how all this happened? The advanced space aliens came to the Earth. How advanced this space aliens are? They can create any animal from parts of other animals; design new creatures and animals; and recreate animals that have been extinct long time ago – l~~l~~ike the dinosaurs. Here they found us – "smarter than the other monkeys." The space aliens calculated that we began our "development" too "late" in the "cycle of life and death of the galaxies," and decided to "help" us get "developed" faster. They gave us more brain power, and devised elaborate plan to give us technological knowledge (when we sleep). The information for our technological advancement has been given to us over some period of time. In the same time, they wanted this "technological advancements" to look as if "we made the necessary progress through our own "discoveries" when in reality most of this knowledge is given to us when we sleep. They have some schedule for advancement of knowledge given to us. Space aliens do not have to come to the bedroom where the "receiver" of the scheduled knowledge has to receive it. Somehow they work through some "balls of light." Such balls have been reported to be present above of some crop fields where crop-formation messages are given. Humans that make FAKE crop-formation "messages" are doing great disservice to the human race – as we are observing real and fake messages simultaneously. Here we have messages from the real space aliens, and human idiots making something to appear like their messages. Can we get more stupid than that, of course, we are humans therefore stupidity, dishonesty, back-stabbing are the main features of our characters. We could have self developed ourselves without space alien interference, but because we are too late in the game, the space aliens decided differently. Accept your fate, and have faith in the "grand design" for us. Instead to be "dust in the wind," let us be CONSCIENTIOUS DUST IN THE WIND! It is not easy to survive in a nuclear reaction after all, but this is our future! Children, be brave as you shall and must prevail!

All these religions are given to us by them. Notice that the entire Earth is covered with religions. So the goal is that every tribe has to have some religion. Obviously this is the decision of the "central committee of the ruling party" of the space aliens. Put yourself in their shoes, it is NOT EASY to develop "true humans" from "smart monkeys." Trough natural evolution we could become true humans; give or take 100,000 years.

Let's look at these predictions of future events. I am reading Lermontov, and he is informing the readers that there would be no king but Lenin. (The name Lenin was not given.) This prediction is made in 1830; that means that the Bolshevik revolution was in the script for the future history of Russia almost 98 years before the actual event. I want to address the Mensheviks (that lost the battle against the Bolsheviks): Do not feel bad that you lost; the space aliens have decided who the winner should be – the Bolsheviks.

Let's get back to the science of Astronomy. Everything that Einstein has said is wrong. The only exception is for what he received Nobel Prize. Anything else is absolute nonsense. I am puzzled, yet our history is prewritten by the space aliens. Why do they allow all this misinformation? And it is not only Einstein's complete nonsense, there are many scientists throughout the history spreading halve-cooked lies; take for example Newton, Eddington, Hubble, Doppler, Hawking and many others. Seems to me that we humans are not without smart people, but somehow the stupidity always manages to get on the top. Probably you have read books from some smart people, but their wisdom is not the wisdom of the collective heard. No matter how good of a system we design, give it some time, and the collective stupidity will destroy it. Is Einstein servant of Satan, or servant of God? If he is servant of Satan – his complete deceptions still reign. If he is servant of God – God is saying to us to put to scrutiny the established idiotic believes, but looks like we are extremely slow learners.

One thing is absolutely certain - we cannot govern ourselves. Throughout the history humans have tried all methods of governance, and the result is always the same – failure. At this moment I will call Einstein servant of Satan, but if we put him as servant of God, it is only to show us how stupid we can get through believing in his absolute nonsense.

All I can do is to inform you about the real universe and the previous and current deceptions the rest is up to you.

WHERE ALL RELIGIONS MIGHT FAIL

We are in uncharted water, and we have found ourselves where CONVENTIONAL THINKING AND CONVENTIONAL RELIGIONS WOULD NOT WORK! Maya predicted that we will enter in a new era 2012, and we are in it right now. People live with the notion that only human activities and the release of too much CO_2 are to blame for the global warming and the weather havoc which is happening right now, but there is another player more powerful and nothing we can do about to stop it – that is the influence of the intruder. I already explained the trouble for our BS, but the question is can the religions help us? I will render my humble opinion. I think that the religions work if the BS is working properly. Religiously, we are in uncharted waters. As much as I hate to give bad predictions, we are heading for trouble; maybe the worst is yet to come, and there is no light at the end of the tunnel yet. I would say: Do not abandon your religion, BUT KNOW THAT NO RELIGION

HAS POWER OVER THE UNIVERSE! ~~universe is the primary power, and our fate depends on it.~~

    -Instead of living in peace and love, we live in the state of war and strife. ~~Yes,~~ Wwe cannot affect the intruder, but we can defend ourselves to certain extend from potential falling asteroids, and store food for longer periods of time in case that something hits the Earth or the intruder ~~and~~ render it "no-crop-producing" for several years. Another thing for sure is climate change lasting (200 years?) which might change global weather patterns which might change river-flows, desert locations, and other unforeseeable weather phenomena. None of the current big scientists understands what is really happening!~~g. Can we try to create some artificial magnetic field? This idea sounds absurd, but that is what we need to know for eventual living on Mars or the Moon. Imagine creating some sort of localized multi-layered protective magnetic field.~~ I read that ~~to the answer why~~ some space aliens left their planets ~~was~~ because something bad happened to them~~ir planet~~. This could eventually happen to ~~ours~~us. At this moment we cannot leave the Earth because we do not have UFOs, and we do not know where to go. The dinosaurs did not know what hit them - we might know, but at this moment we will suffer the same fate.

The scientists with their misunderstanding the gravity of the current situation are talking to you rosy rubbish and eating away your limited time for real salvations. Why not having something sort of insurance, and prepare ourselves to be space aliens? Let me recap, the Earth belongs in the realm of the universe – no gods, space aliens, or humans have any power over it – the universe commands supreme. Keep your eyes pilled, if the space aliens are leaving the Earth, then thing are not too rosy for us. My assessment of the situation is that the human race probably will survive somehow, but we have to do the right thing – be prepared! WHAT YOU DO IF TERE IS NO CROP FOR 30?~~.~~

THE NEW RELIGION

Somewhere in the in the internet was post~~ed~~ed the question, why no new religions have been invented in recent~~ly~~ years? How about the intolerance ~~of~~ from the existing ~~once~~, and the fact that they are government monopolies?~~are sanctioned by some governmental bodies?~~ The message is clear,~~-~~ that people~~we~~ do not need new religions. False ~~, and p~~propaganda also is spread t~~that~~ the ~~existing religions are true and any~~ new religions are dangerous and ~~can be only~~ fake. ~~Any new religion could be only false, and nothing good can come out of it.~~ There should be some tolerance towards~~space for~~ new religions, and the society has to fallow the laws of tolerance. Let the people have a ch~~oice~~ance to choose, and not only some conspirators behind the scene that pull~~ing~~ the strings. Some people think that their opinion should be the opinion of the world. They will use any means available ~~or not available to them~~ to impose their point of view.~~- The "collective wisdom" of~~

~~the society is manipulated behind the scenes. The institutional WISDOM of the governmental mafia is the wisdom of the idiots.~~ We have good principles already written, but <u>no one</u> ~~do we~~ follow<u>s</u> them<u>!</u>~~?~~

I will introduce a new religion, <u>because it is NECESSARY right now. Take this religion as a supplementary. The exiting religions could not get you through these trying times.</u> ~~and let the dice fall where it may.~~ There is some difference between the establish<u>ed</u>~~ment of previous new~~ religions and mine. Many of the prophets have received their messages from God, in my case I have to make the message myself. I should call my religion philosopher's religion. I have to assume that God or the Gods approve of what I am doing, but beyond that whatever.

Let's the new religion be introduced, but before that let me express some pessimism. All is in our honesty, but the human stupidity is fathomless and wide spread. The struggle between good and evil and smart and stupid is forever. Yet somehow we are here, and the "hope" dies last. Let the hope for "the good and the better" prevail<u>s</u>, amen.

Secret societies are the sewer of human thought and action. ANY ORGANIZACION THAT IS CLOCED FROM SCRUTINY BECOMES A SEWER. Anything that concerns the society as a whole has to be in open forum<u>s</u>. Figuratively speaking, when the door is closed for other concerned <u>observers; the conspirators and servants of Satan take over.</u> ~~to see and hear things are to become wrong.~~ This ~~is a~~ rule <u>has</u> never failed<u>s</u>, but the humans never learn from their mistakes. <u>ALL SECRET SOCIETIES ARE OR EVENTUALLY WOULD BE RUN BY SATAN!!!</u>~~Secret societies are run by Satan.~~

SERVING SATAN AND GOD SIMULTANEOUSLY

We humans are capable <u>of,</u> and this is widespread modus operandi<u>,</u> to serve one minute God and the next minute Satan, and most of the time we are not even aware that this is happening. A line has to be drawn between when one serves God and when the same person serves Satan. <u>IF ONE DOES NOT SERVE SATAN; ONE SERVES GOD</u> ~~If you do not serve Satan you serve God~~ – this is the dividing line. And if someone wants to make fun of this rule and say<u>:</u>~~"~~<u>:</u> "I see a person reading a newspaper, is this person serve God at this moment?" And the answer is yes, because the rule is "if he does not serve Satan at <u>certain</u> ~~this very~~ moment<u>,"</u> and we assume that he is not, therefore he is serving God.<u>"</u> And someone might ask, how we can say that he is serving God compared to Mother Teresa. Serving God has different degrees, and certainly Mother Teresa is in <u>the</u> highest~~r~~ degree than the person reading a newspaper. The principles have to be followed to the letter. The evildoers consider themselves righteous. They assume that they are on the side of "good" when the opposite is true. Most of the religions are passive, and these are the religions governments like – prey and say nothing.

Bulgarian church is Christian Orthodox. Once I was listening to what the priest was saying: "Let's pray to God to save us from Satan," then I think I heard a Jewish priest say the same thing, so I am not sure. But, let's get to this phrase, one supposed to pray to someone else to be saved from Satan. I have problem with this. Reason is that I see people serving Satan, then as "good "Christians" they follow the priest advice, and everything looks "normal," but it is not. The element of PERSONAL RESPONSIBILITY and PERSONAL COMMITMENT is missing. I want to hear the phrase: "I will not serve Satan when I find out that I am doing so." Let's say an evildoer enters the church. Supposedly he is servant of Satan. He makes a cross; lights a candle; listens to the priest; prays to God to save him from Satan; and everything looks normal except that he is a servant of Satan himself. Let me make fun of my new religion. I am not sure at this moment would my teachings turn out as a new religion; a new political party; or a moral guidance, but I will forge ahead, and let the chips fall where they may. "The Church of Truth" has to be open, and I am ready to do so. IT TAKES A VILLAGE, HELP THE CAUSE! GLORY TO GOD!

www.ingramcontent.com/pod-product-compliance
Lightning Source LLC
Chambersburg PA
CBHW052359220526
45465CB00003BB/1172